D1272346

A Citizen's Guide to Ecology

A Citizen's Guide to Ecology

Lawrence B. Slobodkin

OXFORD
UNIVERSITY PRESS
2003

OXFORD

Oxford New York
Auckland Bangkok Buenos Aires Cape Town
Chennai Dar es Salaam Delhi Hong Kong Istanbul
Karachi Kolkata Kuala Lumpur Madrid Melbourne
Mexico City Mumbai Nairobi São Paulo Shanghai
Taipei Tokyo Toronto

Published by Oxford University Press, Inc.
198 Madison Avenue, New York, New York 10016

www.oup.com

Library of Congress Cataloging-in-Publication Data
Slobodkin, Lawrence B.
A citizen's guide to ecology /
by Lawrence B. Slobodkin.
p. cm. Includes bibliographical references and index.
ISBN 0-19-516286-2 (cl.) — 0-19-516287-0 (pbk.)
1. Ecology. 2. Nature—Effect of human beings on. I. Title.
QH541.S54 2003 577—dc21 2002072826

987654321

Printed in the United States of America
on recycled, acid-free paper

To Yan, Mathew and Liaht

Contents

Acknowledgments

My wife, Tamara, and daughter, Naomi, put up with me during endless writing and rewriting.

My colleague Manuel Lerdau provided important criticism. I also tried out other ideas on Stony Brook colleagues, particularly Dan Dykhuizen, Mike Bell, Charles Janson, Lev Ginzburg, Jessica Gurevitch, and Geeta Bharathan.

Doug Futuyama, Yossi Loya, Rob Colwell, Uzi Ritte, Rosina Bierbaum, Phil Dustan, Scott Ferson, and Conrad Istock are among my former doctoral students who taught me more than I taught them.

Since 1947, my friend Fred Smith of Woods Hole has provided wisdom.

The late Evelyn Hutchinson of Yale demonstrated to me that ecology is worth a life's effort.

Kirk Jensen has provided important, patient criticism and encouragement well beyond the usual role of an editor.

I have omitted the names of many other people who have been important in my life. A list of who they are and what I learned from each of them would be a thicker book than this one. I ask their indulgence.

A Citizen's Guide to Ecology

Introduction

DEFINING ECOLOGY

Ecology studies interactions among organisms and between organisms and their environment in nature and is also concerned with the effects that organisms have on the inanimate environment. It is concerned with not only what kind of air a species must have but also what effect that species has on the air.

This book is not an elementary ecology textbook. A textbook would be longer and more didactic. Ideally it would present a survey of what is being done by the 7,600 members of the American Ecological Society and their students and collaborators, and it would prepare students for more advanced, specialized books covering one or more of the sixteen subdivisions of the science of ecology that are listed by the society.

This book is simply a description of what happens outdoors today. What has been happening outdoors for the past billion or so years? Has it changed much and is it likely to change further? How do you and I fit into the changes and the constancies?

I have two goals. One is to enhance appreciation of the pleasure and beauty to be found in nature. Another goal is to help individual citizens understand the real and unreal assertions about existing problems and impending disasters in nature.

There is one important difference between ecology and many other fascinating sciences and games: Unsolved problems of chess, astronomy, or mathematics will not change if we ignore them.[1] Our activity or lack of activity can alter the state of ecology. A major focus of ecology is on determining how certain aspects of the natural world change or do not change. We must ask:

- What properties of our environment will stay constant, regardless of what we do?
- What changes are inevitable?
- Are particular changes desirable?

In the first chapter of this book I will present some of the large-scale mechanisms that underlie all of ecological change and constancy. Chapter 2 focuses on individual organisms, species, and landscapes. The third chapter examines how we can reach reasonable conclusions about specific practical problems.

Chapter 1 describes big, inexorable processes that are almost Wagnerian. The second chapter describes smaller, quicker, more complicated, and sometimes almost playful processes that are more like Mozart. Chapter 3 struggles to make sense out of how the material in the first two chapters is used in making decisions.

Motives for the study of ecology range from a sense of awe to a sense of alarm. The study of nature can be a purely intellectual exercise or can focus on practical problems. Fortunately, the world in which ecological problems appear is extremely beautiful, and thinking about ecology can be a great pleasure.

Descriptions of nature can be dramatic. The list of characters includes molecules, mountain ranges, lions, butterflies, real serpents and dragons, and ultimately all of humanity.

There is a temptation to infer mysterious causes for natural events, as if nature had very human qualities such as kindness and wisdom. I will firmly resist this temptation. Reliance on the nonexistent does not help us solve real problems. Nature doesn't care. It simply is.

There is no way to provide a recipe book of correct solutions to all ecological problems. Practical problems must be solved in particular environmental, social, and political contexts. It is, however, relatively easy to provide enough background in ecology so that you can evaluate the issues and the combatants when aspects of your world are at stake.

If you want more information, the references at the end of the book list some of the books and articles I read in preparing the text; you can also read them to check my statements if you want to.

CHANGES

The practical problems of ecology are all concerned with changes. Earthquakes, moving glaciers, years of drought and years of flood, volcanoes, and building projects all cause ecological changes. Sometimes new kinds of organisms burst onto the world scene and make major changes. Even in the absence of human disturbance, the world goes through changes of many kinds and on many scales. The practical concern with ecology is based on the real possibility that we are disturbing the world in dangerous ways and that our understanding and knowledge are so inadequate that we can cause irreparable damage without even noticing until it is too late.

Permanent changes have occurred, and more will occur, but most changes are not catastrophic. If it could be done, I would like to return the ecological condition of the earth back to what it was a few hundred years ago. It cannot be done. Some changes, such as species extinctions, are permanent.

Some changes would never occur without humans. We pave roads, build cities, and blow up mountains. Generally, our excuses for doing these unnatural things come from commerce, industry, agriculture, or military intelligence (an oxymoronic concept). In any case, they are almost never done with the primary purpose of improving the ecological world.

Perhaps cities could be destroyed and roads ripped up and the land they used to occupy could revert to wilderness or farms. Perhaps poisoned rivers could be rehabilitated. These are problems of such a large scale that agencies other than governments are not likely to be effective.

Many changes that are not our fault may suddenly change our world. Earthquakes can flatten cities or change the path of rivers overnight. Volcanic eruptions can create new islands or new mountains or make old islands and mountains disappear in a matter of a day or so.

Recently an ice platform the size of Rhode Island was set afloat at the edge of Antarctica. It quickly broke into floating rafts of ice the size of football fields. The effect of this on animals that normally breed on stationary ice sheets is yet to be assessed. How much did our contributions to global warming precipitate these events? I don't know. Can we repair the situation? Not directly, but perhaps we can use it as a lesson for future behavior. Will we? I doubt it.

Some sudden changes can be less dramatic but perhaps as important. The last members of some species that have been around for scores of millennia may die tonight and the species will be gone forever. This will probably be a bacterial species that has not yet even been discovered, and the effect of its absence will probably be negligible, but I am not at all sure.

A gene may mutate tonight in a virus or bacterium and start a global plague tomorrow. A functioning system of public health and ecological monitoring could lessen the impact of the plague. We do not really have public health facilities for this, and we certainly have nothing like an ecological response system.

Some natural changes are relatively slow. The pin-cherry tree that stood on the edge of a little sandbank last year fell into the river as the bank eroded. The gully in a hillside has deepened and must now be avoided by walkers. In the spring a little stream runs brown with clay particles that it is carrying into an estuary. Many of the big changes that have occurred on the earth are the sum of these little changes, extended over large areas and many centuries.

There are changes on scales of thousands of years in which rocks and sediments erode or are added to. Ultimately, over millions of years there are transfers of minerals from ocean sediments to uplifted rocks and back to seawater solution. None of these changes requires humans, nor can humans prevent any of them.

Although human activity cannot destroy the global ecosystem, we can change it in ways that will be unpleasant for us. Most human activities that have any ecological meaning are massive accelerations of processes that have always occurred, such as flattening of mountains and breaking of rocks.

The earth's atmosphere is particularly sensitive because its natural movements are rapid and the total quantities involved are relatively small. Not only carbon dioxide concentrations can be changed but also the concentrations of other gases, including many different compounds of sulfur, nitrogen, and carbon.

For many millions of years the ancestors of humans played like clever squirrels in the trees and flatlands, changing ever so slowly. For unclear reasons, one primate lineage developed a capacity for intelligence and behavioral complexity that had never been developed before. Our ancestors, the hominids, are distinguishable as rare fossils three million years ago. They became more common three hundred thousand years ago.

For the last approximately fifty thousand years, thoroughly modern humans have lived on earth. Over the past ten thousand years, their effect on the world has become increasingly conspicuous.

If by pristine nature we mean nature free of human influence, there hasn't been any for thousands of years. Almost the entire globe shows the effects of humanity's passage or residence. Forests were burned, stones were massively rearranged, and crops were planted where there had been brush or wild grasses. Water was rerouted and wastes added to rivers. During the ice ages, there were, in North America, wild horses, camels, and elephants. They were probably hunted to extinction by hungry humans.[2]

No part of the world is what it was before there were humans. Forests, grasslands, seas, lakes, and rivers have been, and are being, changed by human activities in ways that many find unpleasant or even dangerous. Desirable and interesting organisms are often found to be declining in numbers, and unwanted organisms are often found to be increasing.

The chemistry of the air in even the most remote parts of the earth is different from what it was just a century ago. Clamshells in the most remote ocean depths have lines of radioactive iodine from the fallout of cold-war atom bomb tests. Lead from industrial smoke is found in the ice of Antarctic glaciers. The chemistry of the atmosphere has changed not only because of furnaces and automobiles and the loss of forests, but also from what would seem to be small things, such as the propellants for hair spray and underarm deodorants. Even the temperature of the earth is changing, at least partially as a result of human activity.

Concern with humanity's effect on nature is at least as old as written documents. Until recently this concern was tied to questions of privilege more than concern with nature itself. Good areas for hunting and fishing belonged to families or tribes.

From the beginnings of kingship in ancient Egypt until today, one privilege of the rich and powerful has been to hunt in game parks, from which less important people were excluded. In twentieth-century Scotland, fishermen purchased the right to fish in streams protected from use by poor people. Now that the wholesale killing of wildlife is no longer fashionable, there are photographic safaris for the wealthy.

Ecological changes by humans go beyond killing or not killing other kinds of animals. For centuries cities smelled of horse dung, sewage, decaying vegetables, and more recently coal gas and kerosene. It was presumably less healthy than clean-smelling air.

The belief in damage to health from foul odors is ancient. Note that the word *malaria* is Latin for "bad air." In fact, before the role of mosquitoes in transmitting the disease was known, staying away from swampy air was the best protection against the disease. The U-bend in the drain of any modern sink or toilet was designed to place trapped water between the foul smells of the sewer and the interior of a house to preserve health. We now have chemical dispensers that add to the gas in a home fresh-smelling perfumes or mild anesthetics to produce nasal numbness and thereby eliminate odors.

During the last half of the twentieth century the departure of horses and wagons removed many of the odors from the streets of most American cities, but the odor of gasoline and diesel fuel remains.

Ecological changes can be very local. Because of the building of roads and houses during my lifetime many of the places I explored as a child are gone or inaccessible. Wildflowers that were common are now rare and unpickable. Fishing is not what it was—and fishermen from the time of the first worm impaler have had the same complaint. Hunters, who relish the past as much as fishermen, find that game has become scarce.

There are sometimes legal mechanisms used to oppose certain kinds of change. Species may be declared to be endangered and must then be given special consideration.

Not everything is changing, and not all changes are bleak. Sometimes surprising organisms persist in unlikely places. Cacti are associated with warm deserts, but there is a population of prickly pear cactus plants growing near my house on Long Island. There are fewer than thirty plants. For more than thirty-five years they have not increased, but neither have they disappeared.

In America eagles, wolves, and beavers are becoming more common. Coyotes may soon become a problem in Massachusetts, and brown bears are worrying New Jersey suburbs. Mockingbirds and cardinals were considered southern birds fifty years ago. They have now invaded Long Island, to everyone's delight.

One change initiated by the U.S. Forest Service in the early twentieth century was to fight the occurrence of wildfires.* This was successful enough that undergrowth and dead wood became thick on the floor of western forests.

In 1987 a wildfire of singular intensity, with the capacity to leap over roads and firebreaks, appeared in Yellowstone Park, scorching at least 20 percent of the park. The land looked dead. Almost immediately a book appeared proclaiming "the end of nature."[3] To claim nature has "died" is silly and nonproductive, even if it sells books. Nature changes. It cannot pass away.

In 2000, another drought year, once again uncontrollable wildfires struck Yellowstone. The area of the forest that had been spared in 1987 now was burning. I was in Yellowstone Park during the fire and saw that the formerly burned-over sections

---

* This centered on the Smokey the Bear campaign—a masterpiece of effective advertising.

were a green tree nursery of small lodgepole pines, unburned by the new fires.

Fires had happened before and will happen again, and the forests of Yellowstone Park will certainly change as a result of those fires, but the changes will not by any means be catastrophic. There will still be forests in Yellowstone Park. In fact, the cones of several of the trees of Yellowstone release their seeds only if they are baked in a fire.

It is important to understand the different meanings of change in ecology. Some changes seem good to some people other changes do not. It is equally important to understand the processes that resist change. Some of us want to drain wetlands to avoid mosquitoes and create real estate, while others wear insect repellant and relish the many kinds of organisms found in wetlands.

What changes can be influenced by human activity? To successfully oppose or promote specific changes in nature there must be an understanding of how nature works. There must also be legal and political machinery in place and working in your favor.

Concern with ecology is necessary. It is not a fad. We can and do change the properties of nature, although the *mechanisms* of ecology do not change, just as the laws of chemistry and physics do not change.

Nature is neither wise nor benign nor malicious. There are no immaterial forces guiding ecological systems, although these are sometimes suggested.[4] Solving practical problems of ecology requires using science and technology in a political, social, and even religious context.

We are, directly or indirectly, actors in nature. Its rules limit us and its dangers challenge us. If ecologists are very successful, they will help maintain the pleasant and livable properties of the world. If not, the world will change in unpleasant ways.

If we completely fail to solve ecological problems, it will involve the deaths of people as well as of other kinds of organisms. Probably the major practical ecological problems will be only partially solved because of deficiencies in science, technology, and most particularly vision. This does not excuse us from attempting to solve them.

After problems of terrorism, nationalism, political corruption, mass starvation, and crime, the problems of ecology are the most serious in today's world. In fact, ecological problems and their solutions may intertwine with these other great problems. The world in which ecological problems appear is extremely beautiful and its complexities are fascinating. Thinking about ecology is necessary. Fortunately, it can be a great pleasure, even in the shadow of an uncertain future.

WHY ANOTHER ECOLOGY BOOK?

Ecology arose out of an amalgamation of two streams of thought, one concerned with humans and the other with the rest of nature.[5]

More than two hundred years ago, in an influential book, Robert Malthus wrote that increasing human population size was endangering the future.[6] This conclusion is frequently rediscovered.[7] It is certainly valid, but it requires careful analysis and qualifications to be useful in any practical way.[8]

Public awareness of ecology in the United States began when the American frontier disappeared. At first the loss of wilderness had poetic and sentimental meaning. For some it meant loss of study areas. It was not immediately seen as a significant danger—as claiming victims. This began to change sixty years ago. The great Oklahoma dust bowl of the 1930s was seen as the first in a series of prospective disasters.[9]

Academic ecology began in 1903 when a group of British

botanists saw themselves as having a separate specialty and formed an organization and began to publish their own journal.[10] My own professional life of fifty years has spanned half its history.

The 1940s and 1950s saw at least a dozen good popular books on applied ecology, but ecology did not engage a mass audience until *Silent Spring* became a best-seller in 1962.[11] This focused on the danger of insecticides to birds. There had been concern for wildlife before, but not presented in such clear prose and with such a sense of urgency. The modern "ecology movement" had begun.

Ecology is in danger of becoming an uncomfortable blend of a science and a passé but still trendy mass movement. Nevertheless, the residual enthusiasts for ecology have a copious supply of study material available. We are bombarded by magazines, news reports, and enough television specials to make even the most beautiful landscape trite. Small bookstores have a section on ecology or environmental crisis just after cookbooks and before the economics section. Most of the popular ecology books are designed to encourage depression or alarm. Bleak predictions of disaster provide pleasurable frissons, once supplied by such ideas as an invasion from Mars and infant damnation. Literally hundreds of books dealing with environmental and ecological problems will appear this year. A few of these will be picture books of remarkable beauty.

But beauty does not sell books fast enough. Therefore, these lovely picture books will probably include a proviso that they are accounts of endangered islands in a sea of advancing environmental degradation.

Most books about ecology are listed as nonfiction, which is not necessarily the same as being demonstrably true. There is also ecological fiction, in the same sense as detective fiction and science fiction.

Fanatic belief in popular tales about ecology has even spawned terrorist movements. One novel, *The Monkey Wrench Gang*, has been used as a manual for ecological terrorists. It describes how to put nails in logs (to damage sawmills), blow up dams, and destroy construction equipment, all in the name of protecting the environment. The ultimate objects of these attacks are ecological villains who are cutting down forests or endangering rare species.[12]

A secret terrorist organization called the Earth Liberation Front (ELF) burned a house in Mt. Sinai, New York, eight miles from my home, to show support for an animal-rights activist who threw a brick through a furrier's store window in Huntington, New York. The group claims $37 million in perpetrated damages, including $12 million in damage to a ski lift in Vail, Colorado, for encroaching on land inhabited by lynx. They also liberated thousands of minks from a fur farm to die in the wild.

There is a curious real ethical question. If pork were no longer eaten, the number of pigs on earth would drastically decline. Also, the total number of minks alive in the world would be seriously reduced if mink coats were no longer worn and the mink farms closed.

The same group credits itself with uprooting an experimental cornfield at the Cold Spring Harbor Laboratory on Long Island.[13] The field was the research area for the Nobel prize winner Barbara McClintock, who had been studying the genetics of corn for more than fifty years. One of the things this book might help clarify is the cluster of moral confusions that motivate groups such as the ELF.

I can guess, in a general way, the contents of most popular ecology books. The dust jacket blurb will proclaim that it is presenting either an original solution or a new perspective for a well-known problem, or increasing awareness of an as yet underemphasized problem. It will also commend its own originality, urgency, and importance. Almost identical words reoccur from dust jacket to dust jacket, except of course for the names of the authors and publishers.

For both fiction and nonfiction, the text begins with a personal adventure of the author or of a fictional hero. This individual was suddenly enlightened about the urgency of ecology—perhaps on safari, perhaps in a Calcutta or Los Angeles rush hour, perhaps on the banks of a polluted or unpolluted stream. In any case, the sudden enlightenment resembles that of Saul walking on the Damascus highway or Paul in a taxi coming into town from an airport in Bombay.[14]

The book then has a brief repetition of elementary ecological theory, without mathematics but perhaps with graphs copied, without references, from pioneering texts of the 1950s and 1960s. There may follow actual data indicating the existence of a specific problem. Then come a few horrifying anecdotes and the identification of the villain or villains who must bear the guilt for the origin of the problem or the failure to have solved it.

Both fictional and nonfictional heroes vary in age and attitude. Politically, they range from Ayn Rand–type reactionaries to dissident anarchists. Heroes include lovable academics, inarticulate cowboys, Che Guevara–style economists, and aboriginal saints wreathed in hallucinogenic herbal smoke.

The villains may be conditions or persons or social forces. Media and the industrial revolution are favorite villains along with a selection from the following: overpopulation, capitalism, socialism, and most recently liberalism. Other possible negative forces are Judeo-Christian theology or the moral decay associated with its decline or absence, which the author may read into Eurocentrism, racism, anthropocentrism, sexism, or eating of red meat.

Villains may also be chosen from a short list of industries

(most often mining, oil, paper, agriculture, lumber, and automobile manufacturing) and organizations (National Association of Manufacturers, Environmental Protection Agency, Bureau of Land Management). Religious denominations, Catholic or Protestant, can also serve as villains if need be. The Native American, Buddhist, or Jewish religious establishments are not popular as villains. Some sophisticated books identify conspiracies involving combinations of villains or even combinations of villains and circumstances.

Many of the books contain important information, and often much of what they say is true. Some of the suggested villains are actually guilty of various malfeasances. For example, energy producers really are among the most important broad-span villains.[15] However, assigning blame is less important than finding practical solutions to well-defined real problems.

In their last chapters formulaic ecology books advocate solutions. These are of varying degrees of simplicity and practicality. Birth control and abortion, organic farming, mariculture (farming the sea), socialism, laissez-faire capitalism, vegetarianism, feminism, unconventional power sources, and religious conversion are favorites. I can imagine circumstances and problems in which each of these really might help.

These books met a powerful public demand, which appears to be waning. Earth Day, an annual celebration of the importance of ecology, was first celebrated with glorious fanfare and press coverage in 1970.* U Thant, secretary general of the United Nations, proclaimed that it should be celebrated annually, on the

---

* "Earth Day—a day to celebrate the natural wonders of our planet, 'to think about Earth's tender seedlings of life' was first proposed by John McConnell in early October 1969 to a few members of the San Francisco Board of Supervisors and other community leaders especially interested in caring for and improving our natural environment. On November 25, 1969, the final day of the UNESCO National Conference, 'Man, and His Environment,' Cynthia

vernal equinox, throughout the world. It was picked up immediately by the youth culture.

The enthusiasm faded with the years, until Earth Day was something like Arbor Day or Flag Day. University officials were trying to make it popular rather than being afraid it would break out into violence. I was the main speaker at Howard University's 1991 Earth Day celebration and spoke to less than ten students in a large auditorium. Ten years later, at my home university, the State University of New York at Stony Brook, the director of the Earth Day celebrations was a monk, chaplain for the Episcopalian students, who had arranged an altar with a symbolic dish of earth and a flower in a vase. The congregation was small, elderly, and earnest.

A television special on Earth Day 2000 had lower viewer ratings than any competing show. Ecology (and also evolution) books are turning up on the remainder tables in front of cut-rate bookstores.

Has the heyday of ecological concern passed? Perhaps, but I contend that ecology is not a mere fad and that ignoring ecological information is as dangerous as naively accepting all the strange things that are said in its name.

To be ecologically aware we must understand that there are real ecological problems that require solutions. We must also have some sense of the kind of people who are concerned with defining and solving real ecological problems.

We know that the chemistry of the earth has been modified during the last half century by human activity. There is a critical shortage of firewood in the places where wood fires are more than an attractive amenity. Also, extinction of natural popula-

---

Wayburn, one of the youth leaders on Mr. McConnell's Earth Day Committee, presented the idea, and showed the Earth Flag during this presentation at the luncheon. Many expressed support for the idea" (http://www.themesh.com/abouted.html, accessed 8/21/02).

tions and species and reduction of species' ranges is occurring on a massive scale. But to go beyond awareness we must understand the scientific, political, and social contexts of possible solutions to the problems.

Ecological concerns permeate politics and literature. There are, around the world, significant political parties with strong ecological positions. They advocate various regulations or bans on particular ecologically dangerous activities such as agriculture and manufacturing. They also develop specific targets for change based on local circumstances. For example, ten years ago there was a campaign by Japanese green organizations to encourage switching to reusable plastic chopsticks instead of wooden ones, as a way to save the rain forests. There is obvious, if limited, merit in the idea.

Ecological concerns are even more complex than a desire to preserve a viable earth. "Green parties" differ widely in their general political agendas, although most tend to be left of center. Some ecological activism is associated with the far right. Malthus, one of the earliest population experts, was sure that we should reject all notions of welfare for the poor because population growth means the inevitable persistence of poverty.[16]

Ecologists often feel called on to make strong political assertions. On the other hand, many opponents demonize ecology as an error, fraud, and sinister plot.

The late Dr. Dixy Lee Ray, gourmet cook, onetime professor of zoology, governor of the state of Washington, and member of President Reagan's cabinet, coauthored a book entitled *Ecological Overkill*.* The catalogue advertisement said in part, "Ray reveals how these environmental extremists are promoting a

---

* Available from Laissez Faire Books, 938 Howard St. #202, San Francisco, CA 94103. Another title from the same catalogue is *Eco-Scam*, by Ronald Bailey.

superstate with enough power to forcibly reduce the population . . . ban all use of cars, electric power and manufacturing processes based on fossil fuels . . . ban agricultural methods that make cheap food available . . . ban the cutting of trees . . . turn developed land back into wilderness. With environmental extremists wired to the White House, it has never been more important to understand this threat to our way of life" (unexplained ellipses in the original).

More genteel misrepresentation of ecology is expressed in magazine and television commercials sponsored by petroleum companies that show deer feeding among oil derricks or fish swimming through the legs of offshore oil rigs. There have been television commercials showing killer whales leaping and sea lions applauding, by clapping with their front flippers, at the news that Exxon has introduced a few double-hulled tankers since the *Exxon Valdez* oil spill. I am sure that fish on occasion do swim under oil rigs and deer do feed near derricks, but there are deeper problems associated with the search for and use of fossil fuels.

Objective proposals for approaching ecological problems do exist but must be evaluated carefully. Among the best on a variety of problems, some of which are not discussed in this book, were produced by the congressional Office of Technology Assessment (OTA). They were written by working groups of scientists in response to questions from members of Congress concerned with preparing ecologically significant legislation and regulation. The fundamental effort that went into preparing those reports makes them still a strong base on which to build serious conclusions.

Since the OTA reports were prepared for Congress, conclusions were stated as carefully as possible. This sometimes resulted in basically unreadable prose. Former Speaker of the House Newt Gingrich eliminated the Office of Technology

Assessment, so that there is no longer a single authoritative information source of ecological information for congressional representatives.

There is an ongoing stream of reports prepared by scientists for reading by scientists, sometimes for the purpose of advancing information and sometimes for the purpose of increasing the flow of funding to specific projects. There are also reports designed for lay people without central responsibilities for environmental decisions. Some are extremely sound; some are oversimplified and overgeneralized. The problems are varied, and all the information we can get must be evaluated and used.

Why should one book about ecology (i.e., this one) be preferred over hundreds of others?

It clarifies in a legible form fears and worries about a complex science. It will enable readers to distinguish serious ecology from mystical visions of nature provided by well-meaning pantheists as well as nonsense mouthed by self-appointed leaders for personal aggrandizement or from a desire to hear their own voices.

Before dealing with specific problems and calls for action, I will present some of the fascinating phenomena and ideas that form the backbone of ecology. Bruce Wallace has made a powerful plea that ecology should be part of a liberal arts curriculum.[17] In one sense I am trying to present an introduction to practical ecology as a liberal art. A good liberal arts course on Shakespeare permits a more sophisticated reading of his sonnets and plays. After finishing this book, the reader should be able to approach other books and articles on ecology in a more sophisticated way.

I hope to look at the underpinnings of ecology—what I see as its significant ideas and contributions but also its implicit assumptions. The popular introductions to ecology may contain intellectual traps, obvious absurdities, and some more-subtle errors. I hope I have pointed out some of them and avoided errors of my own.

How can scientific conclusions be translated into practical action without causing major social damage? I will present what I think must be done to prevent, or at least delay, ecological disasters.

In short, I present as solid a body of ecological information and theory as I can and also criticize some of the dubious material that has appeared in the name of ecology. I hope to provide enough background so that each reader can independently evaluate the strident ecologisms and anti-ecologisms that will be used for preparing legislation, advertising products, and constructing political agendas. How can one identify valid information and reasonable choices of activities while avoiding hysteria and self-serving proclamations? How do we tell the difference between real problems and needless alarm? These are major questions.

Bertrand Russell published a book called *Unpopular Essays*. He explained that the term *popular* had been usurped for material that is obvious to a not-too-bright ten-year-old, so anything requiring more intellectual power than that must be termed *unpopular*.[18] In that sense, this is an unpopular book. The prime excuse for writing an unpopular book is that the subject matter is of critical importance.

The picture I present in this book will not be complete, because there is both too much and too little information. For example, although hundreds of volumes are filled with studies of the ecology of forests or lakes or streams, we still lack sufficient information to answer the full range of specific questions that might be asked about any given lake or forest.*

---

* Even such locations as Wytham Woods, Oxford, studied by Elton and his successors for more than a half century; Lake Mendota, Wisconsin, studied from

Fortunately, it is possible to learn what ecology is and how it is practiced without providing a complete descriptive account of all that we know about organisms. I will try to present some of the ecological processes occurring in nature and some of the traps that are involved in coming to conclusions about ecological problems.

WHO ARE ECOLOGISTS?

To a large degree the definition, popularization, and solutions of ecological problems are in the hands of ecologists. It clarifies the issues to know who these people are and how they are educated.

Where do I stand in ecological battles? I have a parochial interest in my family, and in humanity as a very special species. I also feel that all life, of all kinds, should be considered as sacred, although I reject questions about detailed definition of "the sacred." I am aware that this raises difficulties and produces inconsistencies between belief and behavior. I do not advocate universal vegetarianism. The most consistent of vegetarians are members of the orthodox branch of the Jain sect of India. They wear a veil over the face to avoid inhaling, and thereby damaging, small insects and bacteria.

None of my family members wants a fur coat. I am horrified by those who make coats of the skins of endangered species, but I am not repelled by fur coats in general. I am inconsistent, but I do not rate consistency as a terribly important virtue in the real world.

How did my concern for ecology begin? Becoming an ecologist begins with becoming fascinated with nature. When I was

---

the first decades of the twentieth century; and Hubbard Brook, New Hampshire, a longtime focus of studies of the ecology of chemical nutrients, can only provide limited insights.

nine years old I read the accounts written by Ernest Thompson Seton.* My favorite book was *Wild Animals I Have Known,* which purported to tell about the language of crows, the wisdom of wolves, and loyalty and other wonderful things in other mammals and birds. Years later I read that the great naturalist John Muir said that these were wild animals that *only* Mr. Seton had ever known!

When I was in graduate school the reigning expert on muskrat ecology showed slides of "good" and "bad" muskrat habitat. To me they looked almost identical. When I asked him how he could tell the difference between them he told me, "I think like a muskrat." He was quite serious.[19] Writings of this sort are still being produced, but I don't want to linger on them here.

The realization that organisms and landscapes are beautiful is an important starting point for serious study. It appeals to students on the way to becoming ecologists and helped to keep me in the field all these decades. By the beauty of ecology I mean something more than "Oh my!" stories, although underlying many "Oh my!" stories are important insights.

Personal familiarity with organisms and landscapes leads to love. Anyone who has swum in a coral reef, seen pairs of parrots flying over the Amazon, or watched a herd of elephants coming to a water hole for their morning bath has been very fortunate.

Organisms are fascinating. In fact, each ecologist has in his or her heart some kinds of organisms that are obviously worth endless description and study. A typical weakness of writers about ecology is that they stray into long descriptions of favorite organism or communities or landscapes. I have tried to resist this temptation.

---

* Seton was a romantic figure who deeply loved boys and is primarily significant for helping establish the Boy Scout movement. He was not to be taken as a serious scientist by anyone over the age of eleven.

Voyages to mangroves and corals, to whales and wolves are part of the professional life of ecologists. But exotic places are not absolutely necessary in the search for natural beauty. If you have watched a backyard bird feeder, squirrels stealing from the feeder, a mother cat with a litter of kittens, rabbits and squirrels chasing each other for mysterious reasons, a hunting house cat, or a swimming swan, you know that animals have a beauty that can make your jaw drop.

Plants and some fungi, perhaps more than most animals, can look as exquisite and also behave in exciting ways, including touch-me-nots and impatiens exploding their seed pods, thistles sending out parachutists, or giant puffballs sending off their millions of spores in a brown smoke. This is the stuff of elegant nature shows on television, and these shows may be the best way of gaining experience of large animals and plants while staying indoors in a nonallergenic apartment. As one learns a little more the beauty becomes more intellectual and less graphic.

On the first day of autumn this year, I saw that the porchside web of our garden spider in residence was peppered with transparent wings—the spruce cones had opened and their seeds were spinning in the wind. Thoreau has documented the adventures of the tiny winged seeds of autumn and winter.[20] In their quiet miniature way they are as beautiful as orchids. But, having read Thoreau, I could not trust the spontaneity of my appreciation of the seeds in the spider's web. I can recall to some degree what I felt and saw before I read Thoreau.

To each person there is a set of memories. They may be merely memories of childhood, but in most of these memories there are plants, rocks, animals—even if the place they were seen was and still is a slum. These are irreplaceable aspects of a kind of personal ecology, however difficult they may be to quantify or delineate.

I have a nostalgic affection for the rocky tide pools at the end of Bear Skin Neck, a small peninsula in the town of Rockport, on Cape Ann on the Massachusetts coast. I was eight years old when I first looked into the pools. I saw red and green algae, brown-furred sea anemones, lady crabs, spider crabs, and green crabs. There were sometimes starfish (big yellow and brownish warty ones and small blood-red, almost smooth ones). Usually there were little fishes called cunners (inedibly full of bones), periwinkles moving at a snail's pace over the wet rocks, small hermit crabs bouncing along the bottom in old periwinkle shells, and large hermit crabs parading in moon snail shells.* It was endlessly fascinating and varied.

There were no adults to tell me what I was supposed to be seeing and feeling, in fact no children who shared my fascination, but I watched day after day until darkness, hunger, or the tide drove me back to the dry shore. I would return for the next daylight low tide.

I do not feel nostalgic for any one species but rather for the entire group of organisms, with the sound of the water, with the faint salty smell, and perhaps with the roughness of the granite on my hands and knees. It is in these personal ecological love affairs that communities and landscapes have unity.

I tried to return to those same pools in my sixty-fifth year—too late to bring my children, but trying to convince my wife to share that special world of my childhood. It could not be done. I don't believe that the pools were any less rich and fascinating than they had been. The problem was that someone had noticed that the waves crashed in on slippery rocks, so a child might be hurt if he or she was particularly careless and inept. A concrete and steel barrier had been built that prevented anyone from climbing down onto the rocks.

---

* Later I learned the proper scientific names. I use the common names I learned then, which fit more smoothly into my nostalgia.

There are other bits of landscape that inspire me with loving nostalgia—some in curiously urban contexts. There was, and still is, a grove of very old trees in the center of Bronx Park, near streets where no one dares walk at night. Under the roots of those trees, near the little river, are small pockets of coarse clay, which could be made into objects that would hold their shape for a while after drying. In the gravel near the shore were transparent elvers that had returned from their birthplace in the clear water of the mid-Atlantic to swim up a little polluted river into the heart of the Bronx.

Ecologists are usually initially trained in biology, but some start as geologists, hydrologists, meteorologists, or mathematicians. Ecological problems may focus on water, soil, animals, plants, or even mathematical models. Ecologists tend to divide into plant ecologists, soil ecologists, animal ecologists, mathematical ecologists, and so on. Ecologists may be employed by colleges, government agencies, business, or consulting firms. Some work for advocacy agencies.

Academic ecologists are generally not distinguishable from other professors in appearance or background, except for a slight tendency to affect rural clothing styles. Their concerns range from studies of animal behavior to the effect of soils on crops to global warming and its consequences.

Most academic ecologists proceed with teaching and research, deploring what they see as the deterioration of environmental quality from many sources, but not taking on an activist role. They respond in different ways to the popular interest in their specialty.

An essential duty of academic ecologists is to introduce students to ecology. Arguably most of the children in the world are too poverty-stricken to waste time in contemplating nature. Most American children are urban. Some have had the good fortune to have been at summer camp or to have traveled with their parents. Even for these, camp may focus on water skiing and travel, on churches and hotel lobbies. Their knowledge of nature comes mainly from television and magazine articles.

When young Americans arrive at college, we must provide the experience of real organisms in real contexts in such a way as to initiate real thought. This may require ingenuity if we intend to develop "environmental literacy."[21]

Ideally one should learn about organisms by personal observation outdoors, but large class sizes make fieldwork difficult. I had the problem of teaching ecology to an elementary course of nine hundred general biology students in a suburban university. Searching for an intense experience with real organisms, I arranged for all nine hundred to dissect the guts of termites. This activity was strange enough in itself to engage their apprehensive attention.

Each individual termite's gut was packed full with protozoans of five or six different species. These protozoa are among the most beautiful organisms in the world. Some come surrounded by whiplike extensions on their body wall; others seem to open out into parasols with pulsating ribs. Some have streamers of elongated fur; some are swathed in undulating veils. It was a world not yet made trite by television, high school, or the Internet.

The termite guts were effective, but how could I take more than nine hundred students on a field trip without leaving a mud wallow behind? How does one teach about nature without having the bits of forest on the campus trampled to death by eighteen hundred feet? We did succeed, in part because educated observation can lead to generalizable understandings and focused observations may provide some insight without damaging the observed organisms. The key was to focus on observation and minimize the collection of living specimens. Each student carefully observed ten fallen autumn leaves in order to answer the following questions:

1. Has any one of these leaves fallen from the tree without being partially eaten while it was on the tree?
2. How common is it to find leaves more than half eaten?

The students were then asked to insert knitting needles in the ground and see how many newly shed leaves the needles picked up. From this they could determine, if it was late in the fall and the trees above were almost bare, how many layers of leaves there were when the leaves were still on the trees.[22] This way of measuring the number of leaf layers that were on the trees by determining the layers on the ground has obvious weaknesses related to drifting of leaves and sampling, but, except for leaves too small to pierce with a needle, is independent of leaf size.

Part of the exercise is for the students to consider what can be learned from available information. To what questions might all our numbers be an answer? Why are there almost no intact leaves and also no ribs of completely eaten leaves? Why is the number of leaf layers in our forest always around six? Why isn't it one? Why isn't it twenty? Why isn't there a deep layer of this year's and previous years' leaves lying on the ground? I could have told them the purpose of the exercises. I did not. I wanted them to learn where scientific questions come from. It worked for some of them.

Each student was then required to collect fallen leaves to identify the kinds of trees there were in the forest. With around ten leaves per student, the exercises required only nine thousand dead leaves. Anyone who has raked a backyard will realize that nine thousand leaves will not materially lessen the litter cover of a woodlot.

Is this experience equivalent to a camping trip with Thoreau? Certainly not, but it still permits firsthand understanding of some ecological processes. We had demonstrated that close observation of small things can open intellectual windows with wide views.

The fact that essentially every leaf collected had been eaten in part suggests that all of the leaves were edible to some organisms during at least part of their lives. The fact that almost none of the leaves were eaten down to the stems and veins implied that something stopped the consumption process before it was complete. This might conceivably be that the leaves had rapidly "ripened" in some sense so as to become inedible soon after the herbivores took their first bite. Another possibility is that the meals of the herbivores were interrupted by the animals being attacked by carnivores before they could finish eating.

The fact that the campus forest was green until the leaves were changed by the onset of fall therefore could be taken to imply that the herbivore population feeding on them was in fact kept low by predators.

From the fact that there was no more than three years of leaf fall on the ground, we could infer that the detritovores—the fungi, worms, wood lice, and so on—were finishing up all their potential food and were therefore not prevented by predators from growing up to their food supply.

Observations similar to these are extremely common, implying that there are discernible regularities in ecological systems in nature. More specifically, there is an inference that the food energy supply is limiting to the total nonphotosynthetic part of natural communities.[23]

Scientific argument does not rival the charm of a coral reef, but I think it is intellectually beautiful that at least sometimes the complex ecological world permits itself to be understood by a suitable combination of simple observation and thought.

As we look more deeply into ecology we will find ourselves building our conclusions as much as possible on analysis of data rather than on intuition. We will find that often the chief limit to understanding is finding the right questions. Occasionally intuition is still our best method, and also occasionally it is a serious source of fallacies.

Ecology enhances appreciation of organisms in the way that knowing how to read a score enhances music appreciation. Deeper knowledge helps our appreciation of a drop of water or a pinch of mud or a cow pasture.

At the moment there is a plethora of ecological agencies and commercial firms providing advice to all levels of government and all types of businesses. How to evaluate that advice hinges on knowledge of basic ecology. This book focuses on the common principles and processes that underlie all ecological systems. Its assertions are not under dispute.

Commercial firms and governmental agencies, such as the U.S. Geological Survey, the Bureau of Land Management, lumber companies, and utilities, employ ecologists who are typically assigned to study specific problems. Ecological consultants to business or governmental agencies often work in the context of legal disputes in which ecological arguments are needed, or they may be involved in writing regulations. The issues may be as much technical and legal or even mathematical as they are biological.

In the middle of the twentieth century excellent ecological research was done with cheap and simple equipment and a small number of assistants. A rowboat, a net, an aquarium, or a set of mousetraps and tags could be used to advance the field in critical ways.

In the last two decades the expense of ecological research has increased. Teams of workers, ships, satellites, and massive computational facilities are being seen as necessary. This puts pressure on young scientists to focus on the central problem of money.

If an ecologist is working for a corporation or a government agency, he or she must compete for funding within the organization. Generally, ecologists working for agencies or companies have more funds than those in universities but, in exchange, are more restricted in what they can do, what leads they can follow, and how far from existing practices they can stray. Universities are expected to provide more-original ideas, and perhaps more-dangerous thoughts, some of which may be vital in solving intransigent problems.

University ecologists are in a particularly tight situation for funding. In the standard academic pattern, individual mature ecologists usually focus on one particular problem that has been chosen for the appearance of importance and originality and a proven capacity to attract funds and generate publications. These qualities are required for professional advancement.

In the aggregate the effect of professional ecologists is that of a guerrilla army—fighting for the right cause and trying to do it in the right way but without a central authority or strategy. There are many self-appointed generals. There are many good causes tripping over each other, and in the rush sometimes standards of scientific truth may be ignored.

Scientists are occasionally guilty of "teleological suspension of the ethical," or justifying ethical infractions for the sake of some greater and future good. For example, facts are sometimes stretched or distorted to serve the greater good of maintaining the continuity of research. It is difficult for the nonprofessional reader to recognize these problems from published work. I hope this will clear up as you read on.

Almost all books on ecology predict impending doom and are presented with an air of desperate urgency. Many cheerfully predict inevitable catastrophes, with no indication of how, within a democratic political structure, these might be prevented or even delayed. I hope that their taste for tragedy and disaster will never be satisfied. However, I will try to avoid hysteria and will present some aspects of the story that are not usually emphasized.

I

The Big Picture

WATER AND ENERGY: LIFE'S NECESSITIES

Before we look at any specific ecological systems it will help to understand two fundamental facts that are basic for understanding everything else. These are:

- Water is enormously important for all life and is an extremely peculiar compound.
- Organisms are local accumulations of energy. Energy is crucially important for all organisms.

Life could never have begun if water were not so peculiar. The peculiar properties of water have been seen as a demonstration of divine providence or as an extremely fortunate accident without which life would be impossible.[1]

Water and mercury are the only common inorganic materials that are liquid at the earth's surface temperature. Some compounds, such as alcohol, gasoline, ether, and toluene, are also liquid at the earth's temperature, but these are all organic. They are ultimately products of organisms.*

---

* The rare metal gallium melts at 29°C and might therefore be liquid on a very

Almost all liquids contract as they are made colder and expand as they are warmed. Mercury and alcohol contract as they cool, so they are useful in thermometers. When they freeze solid, they sink to the bottom of the container.

Water contracts on cooling, but only until it reaches 4°C. Continued cooling from 4° to 0° causes water to solidify into ice, but as it freezes it expands. That is why ice cubes float in iced tea or whiskey. Freezing a bottle full of water breaks the bottle. More important, when a lake freezes, the ice layer floats on top, insulating the water beneath the ice. This is why lakes and oceans do not freeze all the way to the bottom. If ice sank, they would freeze solid to the bottom, killing almost all of their organisms. Many lakes would never fully thaw in the spring.

All chemicals used by organisms must first be dissolved in water. Water is a great solvent, so that an enormous number of liquids consist of things dissolved or suspended in water. Milk, blood, whiskey, wine, beer, soup, lemonade, and soy sauce all consist of salts and organic materials in water. Elements such as lead and gold, whose compounds do not readily dissolve in water, are either toxic or of little importance in organisms. Gases such as oxygen and carbon dioxide are taken from the air, but in the process of entering organisms they are dissolved onto wet surfaces such as lung linings and chambers in the undersides of leaves.

Large heat changes are involved in making water melt or freeze. To vaporize water requires 540 calories per gram. This is why drying sweat is so effective at cooling your body. It is also

---

warm day. Some non-natural inorganic alloys such as Wood's metal, made of lead, tin, bismuth, and cadmium, can also be liquid at room temperature. Bromine is liquid at room temperature but almost never found in an uncombined state in nature.

why large bodies of water moderate the temperature of adjacent land. The bitter winters of areas of "continental climate" are due to an absence of nearby large water masses.

Whenever liquid water is unavailable active life stops. Humans can live without food for forty days or more but only a few days without water. In hot places just a few hours without drinking water causes death. Some organisms can survive the absence of water for longer periods, some for less.

In most landscapes, plants grow, animals breed, and dead organic material decomposes whenever liquid water is available. As water drains away or evaporates, life slows down and activities may stop until the next rain or flood. Some seeds and even some animals' eggs can become dormant and wait to sprout or hatch until water is available. There are examples of the stopping of active life of adult organisms in the absence of water and return to activity when water is available again.

During the summer on the tar roof of the old biology building at the University of Michigan in Ann Arbor there was a thin layer of red dust. When this was scraped up we could give a tiny pinch of it on a slide to each freshman biology student. They then added a drop of water and watched through the microscope. In around five minutes some of the dust particles would begin to swell and a few minutes later to move, and then shrug and swim off as rotifers—dust-particle-sized animals complete with a full set of internal organs. I do not know how many years the dust could lie on the roof between wettings and still revive.

There are no absolutely water-free deserts, but certainly there are places where water is scarce. In my discussions of ecology I assume that adequate water is present or that rain will arrive in time or that the organisms are capable of stopping their activities when dry and resuming them when wet. In any high-latitude winter, when water becomes a solid rather than a liquid, most life goes underground or goes to sleep. Even some toads can fall asleep when dried and return to active life when the water comes back.[2]

In very hot deserts there are relatively few kinds of organisms, but those that are present have special properties focusing on water and its storage and conservation. A desert plant one foot high may have roots extending down fifty feet to where water can be found. Beetles in the desert of Namibia drink dew that condenses on their own bodies.

Water is either liquid or ice or vapor. Energy also comes in several forms. For most discussions of ecology we need consider only a few of them. There is energy in sunlight. There is energy bound into chemical bonds of food and in the body tissues of both plants and animals. Energy is required for all movements and for the construction of living tissue.

It has been known since the eighteenth century that when animals use food they consume oxygen, produce carbon dioxide, and use the liberated energy for their life processes. It is as if they were engaged in a slow kind of burning of fuel.[3] This process is called respiration.

An old but comprehensible unit for measuring energy is the calorie. One calorie is the heat required to raise the temperature of a gram of water one degree centigrade. When ecologists or dieticians discuss the caloric value of foods they are concerned with how much energy the food can supply for work and metabolism. This can be measured either by the heat produced by an organism that has eaten the food or by the heat produced by actually burning the food as if it were a fuel.*

In my laboratory, many years ago, we measured the calories in a sample of animals in a bomb calorimeter. A bomb calorimeter is a device for burning a small amount of material in an insu-

---

* The calories used by dieticians are in fact kilocalories and are each equal to 1,000 calories, the unit used by most chemists and physicists.

lated container surrounded by water and measuring the temperature increase in the water. From this we could calculate the caloric value of the burned material.

We found that most animals had a caloric value of around 5,600 calories per ash-free gram, which is essentially the caloric value of proteins. A few had considerably higher caloric value, ranging up to 8,000 or more calories per ash-free gram. Pure fats and oils have approximately 9,000 calories per gram.[4] Plants had around 4,500 calories per ash-free gram.

Animals can obtain energy only by eating it. Green plants and colored bacteria perform photosynthesis, which lets them bind the energy in light into energy-rich molecules. These molecules can be taken apart by the plants themselves or by animals that eat them, releasing the energy for work and growth.

In respiration carbon dioxide is given off and oxygen is taken in. Respiration and photosynthesis are mirror images of each other. It takes exactly as much oxygen to consume a carbohydrate molecule as was produced by the photosynthetic process that produced it.

The rate at which energy is taken in by an organism is usually not equal to the rate at which it is used. A heavy meal increases our store of energy, and a session of strenuous exercise or a long fast decreases it.

For most animals there will be times when food is abundant and times when it is absent. The food, or at least the energy from the food, must be stored during the rich food season to be used when food is absent. Many animals hoard food.

I once found a quart of hickory nuts in a hollow tree in Crotona Park, in the East Bronx. Decades later that park became a center of drugs and gang fights. I would not now dare attempt to find the tree. Perhaps the decay of the neighborhood permits squirrels to collect their hoards without being disturbed by nine-year-olds.

Most animals store excess chemical energy as fat. Fat storage is found in animals about to enter times when feeding is impossible—for example, small birds about to migrate, bears in Alaska in late summer, and squirrels in my backyard in October. It is curious that well-nourished clams, oysters, and mussels, unlike other animals, store a major portion of their excess energy as glycogen, an "animal starch."

Plants store energy primarily as carbohydrates (sugars and starches). Some plants use stems for storage, but many plants store extra energy in big lumpy objects, such as potatoes or carrots.

There are some fat molecules needed in the structure of cells. Also, for whales, seals, walruses, and some other mammals such as bears, fat acts as an insulator as well as an energy store. Some unicellular floating plants use oil droplets as flotation devices. This does not affect the generality that fat and oil are usually energy stores.

Eggs also are typically high in fat, permitting the young to develop to a stage at which they can feed on their own. Plant seeds that germinate underground contain starches and oils to support the young plants until they grow into the light and get energy from the sun. Some seeds, including those of lettuces and carrots, germinate in light. These seeds are tiny and have no appreciable energy stores.

The trunks and branches, the supporting structures of plants, are made of wood, which is energy-rich compared to the supporting skeletons of most animals. Most animal skeletons are made of minerals with very low energy content. It is as if animals could not afford to waste energy in the relatively inert form of skeletal support, but plants had energy to burn.[5]

When organisms have more energy than required for bare survival they may grow or reproduce; if the rate of energy inflow exceeds the rate at which growth and reproduction can use energy, then energy may be stored. Why store fat? Chemically, fats are not better for storage than starches. Fats and oils can turn rancid outside of organisms. For the past four thousand years grass seeds, such as wheat and rice, and starchy plant parts, such as potatoes and turnips, have figured as the great food energy stores of human populations. Stored rice was the unit of wealth of medieval Japan. In the biblical story of Joseph, the storehouses of Pharaoh were stocked with wheat, not oil.

The difference in storage products between plants and animals is not because of biochemical limitations. For most purposes plants do not make fats, but they certainly can. Plant fats and oils are only found in and around fruits and seeds. Olives, avocados, and peanuts are anathema to dieters.

There seems to be a commonsense answer why animals store fat and plants store starch. Think of a fine sailboat. Each piece of the boat has a function, and the overall form of the boat permits integration of function. The appearance of the sailboat is not designed primarily for esthetics, although esthetics is one result of fine boat design. On a serious sailboat there is a dizzying array of ropes, beams, knobs, and hooks. A boat for shallow water will be broad and flat and may have a removable board of some kind, a swinging keel or a leeboard, to permit upwind sailing. A boat for deep water will have a narrower shape and a permanent keel. A racer will be designed differently than a cargo boat. In fact, the anatomy of a boat is neatly tuned to its function.

In a boatlike way, the anatomy of an animal is what it uses to move through the world. An alteration of shape can be a serious cause of failure. This partially explains why biologists since Aristotle have been concerned with detailed anatomy. Older biologists will recall a complex structure of bony levers, tendons, and muscles moving a set of jaws in sea urchins. The working out of the details of this bit of machinery is attributed to Aristotle and it is therefore called "Aristotle's lantern."*

A close friend of mine, a critic and translator of Hungarian literature, was dumbfounded to discover that each of the tiny bones in the head of a carp had a name and a function and that at one stage of my life I had learned them all. Detailed anatomical study is one of the keys to understanding how organisms work.

Many kinds of seeds rely on their exact shape for transport. Those seeds store fat. However, the precise shape of a storage root does not particularly influence the survival of a carrot or potato. These roots store energy in the bulkier form of starches.

I suggest that part of the answer to the question of why animals store fat and plants store starch is that animals store excess energy as fat because fat is the most compact form of energy storage. This avoids disruption in their body shape because of energy storage. Plants generally can store energy as carbohydrates, which are biochemically easier to produce and to utilize.

Notice that flowers, whose shapes are all-important, generally do not store energy at all except for the sugar necessary for nectar production. Also, the sugar in fruits is not usually being stored for immediate benefit of the plants. It is there as an inducement to some animal to eat the fruit and transport the seeds to a sufficient distance from the mother plant.

But recall the exception. Why should clams and oysters not store fat? Mollusks have a shell within which the body hangs, almost like the clapper of a bell. Storage of bulky carbohydrates can occur in the body without deforming the exterior shape.

---

* Aristotle is reputed to have taught a summer course in invertebrate biology on the island of Cos. He might really have been the first to make this anatomical study.

How is the passage of energy among organisms controlled? In most landscapes, where there is water the world looks green. Only in exceptional cases are green plants eaten completely. Herbivores trim the bushes and trees but usually do not kill them. There are falling leaves in autumn that were not completely eaten when they were green. This requires explanation.

Plants have an abundant energy supply from the sun but are limited in their increase by shortages of water and fertilizer and are usually not completely eaten by herbivores. If leaves are completely eaten, it is usually by a new arrival—an alien species. When the Japanese beetle invaded New York sixty years ago they ate rose leaves until only the center vein remained.

Local predators often cannot prey on alien herbivores for the first few years. The foreign herbivores have left their controlling predators behind. If an invading herbivore survives at all, it often does more damage to vegetation than all the native herbivores combined, in spite of the fact that the conditions of the invaded site are new to the invader. This suggests that abundance of local herbivores depends on their food and how well they avoid predators.

If predators hold down local herbivore populations, then most herbivores are expected to be well fed. Conversely, the predators that hold them down are expected to be generally hungry.* Generally, herbivore abundance is limited primarily by predators. Predators are usually limited primarily by food supply. In most places most of the time the bacteria, animals, and fungi that live on dead organic matter clean it all up, so very little leftover organic matter accumulates. Therefore, these eaters of detritus, the detritovores, must, as a group, be limited by their food supply.

---

* Notice that we usually eat meat from mammalian herbivores, while consumption of cats, dogs, and lions is a very special taste, usually involving extended cooking to make their fatless, dry flesh tender.

In short, plants, herbivores, predators, and detritovores interact with each other in such a way that a rough balance is usually maintained. Plants are largely limited by lack of water and minerals, herbivores by severe predation, predators by living food supply, and detritovores by the availability of dead organic matter.

This simple plan contributes to explanations of why plants have energy-rich skeletons while animals have mineral skeletons, and why invading herbivores (if they survive at all) are more destructive than native herbivores. This scheme is important enough so that some people are still furious about it more than forty years after its initial publication.*

There are various departures from the general scheme. Some animals, such as bears, skunks, raccoons, pigs, and people, eat both meat and vegetables. If there are high-level predators—that is, predators that eat other predators—this may change the pattern.

The general idea can be useful in managing ecosystems. For example, lakes sometimes have so many small floating plants that their water is green and opaque. Microscopic crustacea (zooplankton) can eat down the phytoplankton, clearing the water, but the water will turn green again if fish that eat the zooplankton are introduced. Managers can add predatory fish to eat the fish that were eating the zooplankton that eat the phytoplankton that make the water turn green, and so on. Sometimes this kind of management doesn't work. For example, an introduced species may not eat what it's supposed to.

The insights about fat and about the relation between predator control and herbivore effectiveness are not revolutionary, nor is the method of attaining them particularly novel. Insight can

* Personal communication, N. Hairston Sr., 1999. The overall scheme is controversial to some degree but is still being taught and used. More than one thousand research papers have referred to these ideas.

proceed directly from observation and common sense. No deep metaphysical, holistic, or emergent properties are needed to understand or to manage ecological systems, although the management process may not be easy, nor is it always successful.

THE ORIGIN OF LIFE AND OF ATMOSPHERIC OXYGEN

For the past billion years the major innovations in the chemistry of the earth's surface and atmosphere were due to evolving organisms.

Many globally significant properties hinge on relatively small differences in the chemistry, temperature, and distribution pattern of the great air and water masses that cover the earth. For example, carbon dioxide is a quantitatively minor fraction of the atmosphere—around 3/100 of 1 percent, or three parts in ten thousand. The total change of carbon dioxide in the last century due to the activities of humans is around 30 percent, around four parts in a hundred thousand. It is generally conceded that this tiny change of carbon dioxide concentration in the atmosphere is an important factor in changes in climate, glacial melting rates, weather patterns, sea level, and more.

The fact that tiny changes can have a major impact because their effects are multiplied by circumstances and events is one of the most intriguing and frightening aspects of ecology. Assessing the effects of prospective changes is the focus of a large proportion of ecological disputes.

Change in the concentration of atmospheric carbon dioxide seems a small problem when compared with some changes that have been brought about by organisms. One of the greatest changes organisms have brought about in the earth's atmosphere is the oxygen-rich atmosphere. Approximately 20 percent of the air is oxygen. If life had not appeared, there would be almost no atmospheric oxygen.

Our fragments of knowledge about the origin of life are sufficient that we need not invoke supernatural intervention or arrivals from distant stars. On the other hand, our knowledge is so fragmentary that speculations are rampant.[6] Many of the attributes of life, singly and sometimes in fairly complex combinations, can be found occurring without the presence of organisms.[7] Although ingenious laboratory experiments with simple chemicals can produce remarkably lifelike chemistry, the origin of life is not well understood. Life originated on earth by some extremely unlikely concatenation of circumstances.

It might be easier to study the origin of life if it were not so ubiquitous. It is not easy to find places that are free of organisms and of the consequences of their activity. Fresh lava, our moon, Mars, and meteorites are the only natural examples of life-free places I can think of that have been actually examined.

In hot-water environments such as the geysers and pools of water or mud in Yellowstone National Park live red, purple, brown, and yellow bacteria. Hot springs may have been significant environments for the origin of life, and some of the remarkable organisms in modern hot springs may be similar to some early kinds of life, but we really do not know exactly how life originated.[8]

Carbon compounds are found in some of today's meteorites, so there may have been some carbon compounds around to be used as an energy source by the first organisms, but this weak organic soup would soon have been exhausted. The first organisms to develop the ability to use light energy to produce higher-molecular-weight carbon compounds from carbon dioxide would have had an enormous advantage, if only in their ability to be active after sunset.*

---

* It seems clear that the first photosynthetic organisms were like bacteria and that the first plantlike organisms were a symbiosis between photosynthetic

I assume that the first photosynthesizers used light energy to produce organic compounds that supplemented their general diet of dissolved organic carbon. Then they must have very quickly evolved to produce all the organic carbon compounds they needed for their own respiratory needs.

Is it obvious why they continued to perfect and enlarge the process so that they could make more organic compounds than they themselves required? Modern plants in nature generally make at least three times more organic compounds than they need for their own respiration. Some of this excess is used for growth and reproduction.

Some early organisms, similar to modern protozoa, evolved into clumps of cells with coordinated activity, becoming "metazoans" or animals. Many of these metazoans contained green bacteria or algae in their bodies to at least supplement their feeding.

Organisms need not maintain a capacity for photosynthesis. The first herbivores must have been a strong impetus to having photosynthesis exceed respiration in plants. As soon as other organisms caught on to eating them, the photosynthesizers had to produce enough to keep up with the amount taken from them by the herbivores.

When photosynthetic organisms first appeared, around three billion or four billion years ago, they produced large, energy-rich, carbon-containing molecules, but they also produced oxygen that could be used in respiration.* For many modern organisms free oxygen is a necessity, but the first organisms did not

---

and nonphotosynthetic organisms. This is important as an evolutionary problem but is not critical for our present discussion.

* Estimates of ages of ancient events change without notice. There is a pleasure in dealing so casually with a billion years, and it is also the safest estimate.

require free oxygen. Even now some of the ecologically most important organisms cannot tolerate free oxygen.*

Evolving a capacity to eat a living or dead neighbor identifies an organism as a decomposer of complex molecules—either an animal, a bacterium, or a mold. Another alternative is to swallow but not kill microscopic photosynthesizers, so that you can almost live like a plant. There are still organisms, such as the dinoflagellate protozoans, that sometimes carry photosynthetic symbionts and act as plants and, as circumstances change, discard their photosynthetic symbionts and live strictly as animals.

Green hydra can live for months without animal food, relying on the unicellular algae that live within their gut cells. There is, however, a danger that the symbiotic algae can act as parasites, destroying their hosts.[9]

All reef-building corals contain symbiotic dinoflagellates that supply them with oxygen and carbohydrates. Some eat zooplankton as a supplement.[10]

The giant clam, *Tridacna,* figured in old movies about coral reefs, grasping divers' feet. That may have happened once or twice, but more interesting is the fact that the soft tissue that lines the gaping shell is loaded with an enormously heavy population of symbiotic algae, which are the complete nutritional suppliers for the clams. The giant clam may be the largest animal to live only on symbionts, without taking in any solid food. Most animals must rely on eating other organisms to stay alive.

The fact that photosynthesis produces oxygen does not explain the 20 percent concentration of oxygen in the atmosphere. Photosynthesis is the precise mirror image of respiration.† Where photosynthesis uses carbon dioxide, captures

---

* All of the bacteria able to take nitrogen from the air and make it into nitrates that can be used by plants are unable to tolerate free oxygen.

† Glucose and molecular oxygen are produced from carbon dioxide and water,

energy, and produces oxygen, respiration uses oxygen, releases energy, and produces carbon dioxide. Essentially all the photosynthetic products are respired away.

Plants respire around a third of the energy they fix in photosynthesis for their own use. Herbivores eat a large share of living leaves. Almost all organic material that falls to the ground, as dead leaves and wood, is eaten by decomposers—bacteria and molds. Some organic material may be consumed by wildfires. In these processes essentially all of the oxygen produced by photosynthesis is recombined with carbon to produce carbon dioxide. Where did the atmospheric oxygen come from? Thinking of it in a slightly different way: If respiration and fire are the precise reverse of photosynthesis and if all organisms, plants and animals, use respiration, why was any oxygen left over to help form the atmosphere?

Part of the explanation is the important concept that things in nature never come out quite even unless there is some mechanism that forces them to. Over global space and geologic time very slight failures to come out even result in massive accumulations or eliminations.

When a withered leaf falls it will probably rot away completely and all its carbon will become carbon dioxide. All the oxygen produced in the photosynthetic process that made the

—

using light energy and a massive amount of biochemical and cytological machinery. For present purposes the detailed machinery can be ignored. Conversely, glucose can be burned or respired by appropriate biochemical machinery producing carbon dioxide and water and in the process supplying energy for all kinds of living activities. The process can be summarized by a chemical equation:

$$C_6H_{12}O_6 + 6O_2 \rightleftharpoons 6CO_2 + 6H_2O.$$

It is important to notice that the process of respiration, or simply burning in the presence of oxygen, reverses the chemical equation precisely, producing carbohydrates from carbon dioxide and water.

leaf will be respired away. But what if the leaf falls in a dry desert, where things cannot decompose, or is buried in the bottom of a lake with no oxygen in its deep water? Then the oxygen released in the making of the leaf is left behind in the atmosphere without being able to recombine with its carbon.

Why doesn't the oxygen combine with some other carbon, or, since we know that oxygen is a highly active element, why doesn't the oxygen combine with something else? Why should free oxygen persist?

We can understand the persistence of atmospheric oxygen only by considering the early history of the earth. Before the atmosphere was loaded with oxygen there was a great deal of unoxidized metal about, mostly as ferrous iron dissolved in the oceans. The first molecules of oxygen produced by photosynthesis combined with these ions to produce oxidation products, the most important being the inert, insoluble, unsightly ferric ion oxide called rust. The red iron ores of the Mesabi Range were formed at that time. Only when the world's surface had completely rusted and the outpouring of oxygen and the burial of carbon persisted could oxygen accumulate in the atmosphere.[11]

Every year most of the photosynthetically produced oxygen was respired, but every year some carbon was buried, out of contact with oxygen, and because the world's surface was very nearly completely oxidized, there was no place for the leftover oxygen to go. It remains in the atmosphere.*

For every molecule of oxygen in our atmosphere there is an atom of carbon buried somewhere that did not manage to recombine. These carbon molecules can recombine to form carbon dioxide and

---

* For simplicity I am ignoring the loss of atmospheric oxygen to newly eroded sediments and volcanic rocks.

remove an oxygen molecule from the atmosphere whenever the opportunity arises.

All of our fossil fuels (oil, gas, and coal) are buried organic carbon. When we burn these fossil fuels the long-delayed reunion with oxygen is finally completed.

There is enough buried carbon to combine with all of the oxygen in the atmosphere several times over. If photosynthesis suddenly stopped, it would take the processes of respiration and burning approximately five hundred years to consume all the atmosphere's oxygen. Of course, there is no way even modern humanity can stop photosynthesis, so this is of interest but not a problem. The oxygen content of the atmosphere has stayed reasonably constant at around 20 percent, and certainly not higher than 26 percent, for at least forty-five million years.

A fair idea of the chemicals involved in living organisms can be gleaned by reading the label on a bag of inorganic fertilizer. These are the ingredients that are vital for life and that the abundance of life has forced into short supply in soil and natural waters.

When quantities of chemicals measured in the world are apparently not changing, it might be because they are inert, both biologically and chemically. Some atoms are too big, too insoluble, or too chemically inert to be biologically useful. These either persist without having much role in biology or are toxic to most organisms. Gold, silver, lead, and mercury are examples. The so-called noble gases—like argon, xenon, krypton, and neon—do not enter biological processes, thus maintaining their inert "nobility."

When a particular chemical does play an active role and still seems to be constant in quantity from year to year, eon to eon, then the rates of its removal and addition must be just equal. This might be a remarkable coincidence. More likely there are some mechanisms that maintain the relative constancy.

There are several mechanisms of control for atmospheric oxygen concentration. Some operate when the concentration is too high, and the others operate when the concentration is too low.

Higher oxygen concentration increases the incidence of wildfires.* The burning of forests removes oxygen from the air.

Higher oxygen concentration in the air also increases the amount of oxygen dissolved in the deep water of lakes and oceans, enhancing their ability to decompose sinking organic matter. Decomposition of organic matter consumes oxygen and produces carbon dioxide.

Conversely, any lowering of oxygen concentration decreases the rate at which material will combine with oxygen. Burning is less likely and decomposition of material in deep water is more difficult. This results in greater burial of undigested carbon.[12] The actual concentration that we find results from these various processes.

There have been other big changes. Each successful new evolutionary mechanism tends to drain the seas and soils of whatever chemical elements are needed to produce the innovation. For example, the earliest seas were rich in dissolved silica. When the microscopic algae called diatoms started using silica to make a glassy protective shell, they essentially stripped the seawater of silica.

When worms first developed hard jaws they established a premium on developing hard shells. Silica was one way to make a shell, but most shelled organisms evolved after the diatoms had cornered the silica market. Most shells are made of some combination of calcium carbonate, a chalky rocklike material, and chitin, a stiffened polymer of sugars and proteins.

The process of evolution is an ongoing source of global

---

* Recall the junior high school demonstration of burning steel wool in high oxygen concentrations.

changes. Evolving animals and plants shifted survival conditions for each other. Algae grew on wet rocks two billion years ago and were followed onto land by worms of various kinds. The algae improved their ability to stand the dryness of the land and finally became so good at it that they crowded each other. When plants are crowded they must grow up to reach light, at first producing the strange forests, known only from fossils, consisting of simple tendrils and mosslike tufts. Later, forests of ferns and cycad palms appeared. None of these have true flowers. It was many millions of years before the first colored flowers appeared.

The point of a colored flower is to attract pollinators and seed dispersers. (Conversely, in one sense the central ecological role of an insect is to disperse seeds and move pollen.) Modern-looking flowering plants appeared only around 120 million years ago.[13]

These are a few facts and speculations about billions of years of the earth's history, in a few pages. The account is not wrong but is not detailed enough to see specific lineages or mechanisms. It suffices to demonstrate that change has been occurring since the beginning. Sometimes the changes have been slow and inconspicuous until some critical point was reached, and then they suddenly burst upon the world. This would have been true of the advent of oxygen in the atmosphere. The rusting of the world may have taken a billion years, with atmospheric oxygen at no higher than 1 percent all that time, until the last massive iron exposure had rusted over. I guess that oxygen concentration would have then risen to something close to its present value within a relatively short span of time—perhaps a mere few million years or even less.[14]

The appearance of constancy in the ecological world depends on the scale of observation and on controlling mechanisms, which often are not very conspicuous at all.

BIG SYSTEMS

Lakes

The entire earth, with its atmosphere and oceans, can be thought of as a single ecosystem. This has the advantage of providing an ecosystem that is essentially isolated. Energy may enter as sunlight and leave as heat, but the system is closed to entering or leaving matter.* Different kinds of ecological systems have been named and studied. These include forests, swamps, coral reefs, deserts, and many more. Many of these have uncertain boundaries, which makes it difficult to describe them completely.

The entire earth is too large for almost all management purposes. We can find out things about the different, not quite perfect ecosystems on earth if we bear in mind that their borders are fuzzy and leaky.

Lakes and ponds usually have relatively clear boundaries. This helps in describing them. They form when water fills hollows in the landscape. The origins of lakes are as varied as the origins of hollows.

Most lakes appear fairly suddenly and are short-lived (decades or centuries rather than millennia).[15] Landslides and earthquakes dam streams and rivers. People dig farm ponds. Volcanic eruptions leave basins.

Thousands of little ponds are produced by beavers and may last as little as one year or as much as a century. In desert areas of South Africa broad, shallow lakes, called playas, appear as a

---

* Material from meteors and comets does land on the earth and light gas molecules do leave the earth, but in small quantities. Also, when giant meteors have impacted the earth they may have "splashed" rocks out into space, and sufficiently violent volcanic eruptions may send material into space.

result of enormous herds of mammals wallowing in slightly muddy spots and each carrying off, caked on its hide, around a half pound of mud.

People can make swimming holes, duck ponds, and water gardens. Damming of major rivers has produced important new lakes. In southwestern England are the dew ponds, formed when hollows in the ground were coated with clay linings, preventing water from dews and light rains from sinking into the soil.

Lake Baikal, in Siberia, and the great Rift Valley lakes of Africa fill faults in the earth's crust. Lake Tanganyika is around five million years old and has accumulated more than a mile-deep layer of sediment in its bottom. The Great Lakes of North America are due to glaciers and are no more than fifteen thousand years old.

Lakes disappear if they dry up or if they become so filled with solid material that there is no longer free liquid in them. Sometimes their outflows are so lowered by erosion that they become merely portions of rivers.

Often rain will produce temporary pools. If they last for even a few weeks, they will be colonized by a surprising diversity of organisms. I saw algae, insects, and toad tadpoles appear within a week in three-inch-deep roadside puddles in West Virginia.

In the 1930s there were quarries in Cape Ann, Massachusetts, producing hand-finished granite paving blocks. They were deep holes in the ground that had filled with water and become swimming holes by the mid-1940s.

Lakes come in endless varieties, but it is hard to live intimately with any lake without falling in love. One of the greatest lovers of lakes was Henry David Thoreau.

Thoreau walked out of what he saw as the corruption in the village of Concord, Massachusetts, and went a full mile away to Walden Pond on a philosophical search. He saw a model for human accord in the "lawful and repeated events" that occur in lakes.*

At the pond he built a very small cabin, planted a vegetable garden, fished, wrote, and watched the pond closely through the seasons. He noted the times of events in the pond, the presence of birds and mammals on the pond's surface and margins, the leaves falling onto the pond, and even the sounds of freezing and thawing. His book *Walden, or Life in the Woods* is an elegant personal statement.[16] To most of his readers Thoreau is of greater significance than the lake, but his book contains very good, careful observations mixed with his philosophical speculations.

Thoreau recorded what could be seen from a rowboat or by fishing or sitting on the shore. It gives an excellent picture. Here is his description of the pond:

> There have been caught in Walden pickerel, one weighing seven pounds—to say nothing of another which carried off a reel with great velocity, which the fisherman safely set down at eight pounds because he did not see him—perch and pouts, some of each weighing over two pounds, shiners, chivins or roach (*Leuciscus pulchellus*), a very few breams (*Pomotis obesus*) (9), and a couple of eels, one weighing four pounds—I am thus particular because the weight of a fish is commonly its only title to fame, and these are the only eels I have heard of here;—also, I have a faint recollection of a little fish some five inches long, with silvery sides and a greenish back, somewhat dace-like in its character, which I mention here chiefly to link my facts to fable. Nevertheless, this pond is not very fertile in fish.

---

* Seeing philosophical and even religious implications in ecology was relatively common until the late twentieth century.

Its pickerel, though not abundant, are its chief boast. I have seen at one time lying on the ice pickerel of at least three different kinds: a long and shallow one, steel-colored, most like those caught in the river; a bright golden kind, with greenish reflections and remarkably deep, which is the most common here; and another, golden-colored, and shaped like the last, but peppered on the sides with small dark brown or black spots, intermixed with a few faint blood-red ones, very much like a trout. The specific name *reticulatus* would not apply to this; it should be *guttatus* rather. These are all very firm fish, and weigh more than their size promises. The shiners, pouts, and perch also, and indeed all the fishes which inhabit this pond, are much cleaner, handsomer, and firmer-fleshed than those in the river and most other ponds, as the water is purer, and they can easily be distinguished from them. Probably many ichthyologists would make new varieties of some of them. There are also a clean race of frogs and tortoises, and a few mussels in it; muskrats and minks leave their traces about it, and occasionally a traveling mud-turtle visits it. Sometimes, when I pushed off my boat in the morning, I disturbed a great mud-turtle which had secreted himself under the boat in the night. Ducks and geese frequent it in the spring and fall, the white-bellied swallows (*Hirundo bicolor*) skim over it, kingfishers dart away from its coves, and the peetweets (*Totanus macularius*) "teeter" along its stony shores all summer. I have sometimes disturbed a fish hawk sitting on a white pine over the water; but I doubt if it is ever profaned by the wing of a gull, like Fair Haven. At most, it tolerates one annual loon.

These are all the animals of consequence which frequent it now.

He had established a kind of communion with an almost self-contained world. His observations are vivid and valid, but he could only access the lake with a coarse net, bucket, and fishing line. Most of what occurred was hidden in the depths of the water. Some of the organisms may have been too small to interest him. Lakes also contain worms, leeches, polyps, mites, spiders, and snails.

Because he had no microscope, he missed all the tiny floating plants or phytoplankton. There are also numerous kinds of bacteria, many which have not yet been properly collected or classified. In almost any lake or pond there are also dozens of species of tiny swimming animals, the zooplankton. These feed on phytoplankton and bacteria, and sometimes on each other.

Some of the organisms in lakes arrive in the water of the inflowing streams. Others occur in lakes but not in streams. Darwin suggested that small organisms were carried into new ponds and from pond to pond on the mud on birds' feet.[17] Basset Maguire washed the feet of birds and raccoons and confirmed their important role in populating ponds.[18] Organisms vary in their capacity to be carried from lake to lake. While many of the smaller ones seem to be dropped onto lakes by bird or wind transport, others are essentially fixed in their locations.

What we find in a lake depends on the properties of the lake as well as on its history. For example, there are ponds all over the world that contain water only in the wet season. There are very small crustaceans and rotifers found in these ponds that grow rapidly, are strongly susceptible to predation by fishes, and have eggs that can withstand drying. One successful reproductive season can establish recurrent annual reintroductions from dormant eggs for at least 350 years![19]

Larger animals must also either abandon drying ponds or hide in the mud. Toads can burrow into bottom mud, grow multiple layers of external skin (like an onion), and sleep until the water returns.[20] Toads also sleep in the superficially frozen soil of my tomato patch during the winter, and I have awakened them with my spade in the spring.

During a drought in Kenya I saw where hippopotamuses had lain side by side in vanishing ponds until the mud had the stiffness of oatmeal. They then emerged, leaving the massive imprint of their bodies in the hard mud. They presumably had gone to seek water elsewhere.

Distinctive new species may evolve in old lakes. In the great Rift Valley lakes of East Africa there occur hundreds of species of mouth-breeding cichlid fish (though the number of species is now being reduced because of the Nile perch, introduced to support fishing).[21] In ancient Lake Baikal, in Siberia, there are hundreds of species of snails and amphipods that are found nowhere else on earth. It even has its own species of seal. The Lake Baikal seal evolved in the lake and is not likely to appear anywhere else in the foreseeable future.[22]

In a lake or pond plants are growing, animals are feeding and molds and bacteria are decomposing the leftovers.

Some zooplankton are so small that they catch single-celled algae one at a time and drink out their contents, as if they were nursing from bottles. When the size discrepancy between animals and their food supply is sufficiently great, a more effective feeding method than catching them one at a time is needed. Cows do not eat grass one blade at a time!

Many animals that feed on food organisms much smaller than themselves use sieves to pick up many at one time, like taking nuts by the handful at bars. Sieving mechanisms differ. Flamingoes, for example, live on zooplankton. The birds scoop up mouthfuls of water. When they turn their heads water drains through the ridges between the flat upper bill and the spoon-shaped lower bill, leaving a mouthful of the zooplankton behind. Baleen whales fill their mouth with water and then squeeze it out the sides of their mouth through a filter of stiff fibers that hang from their hard palate. They can filter many pounds of tiny shrimp out of the two tons of water in each mouthful.

Most zooplankton feed on algal and bacterial cells. These tiny food organisms are as small compared to a zooplankter as zooplankters are compared with a flamingo or as a small shrimp compared with a whale. Many of the small crustacea of lakes and oceans have wonderful systems of spines on their waving legs. These permit them to sift the tiny phytoplankton or bacterial cells out of the water and eat them as pellets, or entangle them in a moving belt of mucus entering their mouths. Details vary from species to species. In many species of zooplankton, the legs are thin-skinned enough so that while the legs are filtering food they also act as gills.[23]

Also in the water are animals that are so small that they do not need gills. Many of these, such as protozoa and rotifers, have hairlike whips on their surface, the cilia, which provide locomotion and can filter bacteria out of the water. The bacteria are tiny compared to phytoplankton.

Some of these small organisms are predators. There are also larger predators, such as small fishes, insects, and worms. These are preyed upon by larger predators, including bigger fish, frogs, salamanders, turtles, snakes, and birds. The interactions among all of these organisms are like the goings-on in a species-rich tropical forest, but they are happening in ponds and lakes all over the world.

For most practical purposes we do not have to study each species in detail nor must we list all of the species present in order to understand a lake. Nor do we have to study each lake as if it were a new world to generally understand what is occurring in it. For many purposes a coarse descriptive level is sufficient. Sometimes greater detail will be needed.

Phytoplankton, and most colored bacteria, can use light

energy to manufacture energy-rich molecules and are called autotrophs (self-feeders), phototrophs (feeders on light), or simply producers (makers of chemical energy from light). All the other organisms in the lake, the heterotrophs (feeders on others) or consumers, have lost the capacity to make high-energy molecules using light energy. Some of these feed on the autotrophs directly, as herbivores, or indirectly, as carnivores, or as decomposers of dead organic material. Others feed on the bacteria that live by decomposing organic material.

Each lake has a watershed, a region of land from which all rainwater and streams drain into the lake. Silt, small stones, and all kinds of organic garbage sweep into the lake from its watershed. The silt and small stones accumulate on the lake bottom as sediment, helping to fill the basin. The organic material is food for molds and bacteria.

Rooted vegetation growing around the shore also adds leaves and twigs to the lake. Sometimes beavers or erosion drops entire trees into lakes. These large pieces of organic material are food for bacteria. Some of the bacteria are fed on by protozoa, worms, and rotifers, which become food for zooplankton that may be eaten by fish.

How much of the organic material that feeds the organisms in a lake is from the drainage basin and how much is from plants floating in the lake water? What is the relative importance of bacteria and phytoplankton in supporting the zooplankton? These questions are still unanswered.

There are also organisms that feed on but do not kill their prey. These are the parasites. Parasites must coordinate their biology with that of their hosts, avoiding prematurely killing their hosts and managing to move from host to host as necessary. This results in phenomenally complex biology. It is the subject of active current research. The field has been beautifully summarized in a popular book by Zimmer.[24]

We have focused on what Thoreau might have seen in his lake if he had a microscope and time. It is an unfamiliar world but is essentially analogous to accounts of the natural history of forests and fields. We must now turn to how the entire assemblage of organisms manages to persist in time.

Lakes Through the Seasons

Many of the ecological processes occurring in watery ecosystems are the same as those that occur on dry land. I think that they are easier to understand in lakes and oceans than in forests. I will therefore begin with the wet ecosystems and then utilize an understanding of wet ecology to help us examine the ecology of dry land.

There are two nonbiological ideas that are basic to understanding the processes occurring in lakes and oceans. One is the idea of solutions of gas in liquids, and the other is the idea of fluid density.

The first chemical formula most people learn is that of water. Water, H_2O, is made of hydrogen and oxygen. The oxygen that is chemically bound to hydrogen to make water is not available to respiring organisms. With oxygen making up 20 percent of the air, land-dwelling organisms usually have no problem of oxygen shortage unless they are in an enclosed space. The situation is different for those that live in water. There isn't the essentially infinite supply of oxygen that there is in the air. Fish must use oxygen gas that is dissolved in the water. Oxygen shortage is a real possibility. This is easy to demonstrate if you are willing to worry a goldfish.

Place a goldfish in a beaker of water. It should appear comfortable. Now remove the fish to another bowl and heat the water in the beaker (without the fish). Bubbles will appear on the beaker wall, and then the water will boil. Now cool it without stirring it. When it is at the temperature at which you started, return the goldfish to the beaker. The fish will exhibit signs of asphyxiation, gulping air at the surface and perhaps being unable to swim. It would be an act of kindness to return the goldfish to fresh unboiled water. It will recover quickly with no harm done. If you vigorously shake the cooled boiled water in the beaker before you put in the goldfish, the fish will show no signs of discomfort.

Heating does not change the chemical properties of the water, but it does bubble off the dissolved air. Shaking the water permits air to dissolve in it.

The second prerequisite concept is that of liquids floating on other, denser liquids. Density differences between water masses are critical to the movement of water in lakes and oceans.

In the days when alcohol was a favored form of sin, there was a fancy bar drink known as a vermouth cassis. It was considered an effeminate drink, whatever that meant, and a test of the skill of the bartender. It consisted of as many as five different colored beverages layered into the same glass.

Liqueurs and sweet wines differ in their sugar and alcohol content. Alcohol is less dense than water, but sugar solution is denser than water. A good bartender would carefully pour them into the glass so that the denser liquids were beneath the less dense. In practice this meant pouring the sweetest low-alcohol liqueurs in first. The idea was to make a pretty drink that would not homogenize readily.

The warming of the top layer of a lake while the bottom water stays cold simply floats less dense water on denser water. In oceans both temperature and salinity determine the density, so that low-salinity water will float on higher-salinity water of the same temperature but may sink under much colder water with lower salinity.

With these two concepts in place we can turn to the workings of a lake.

The phytoplankton in a lake use the minerals dissolved in the water, particularly the phosphorus and nitrogen salts, as fertilizer for growth. The chemical composition of an ideal water bath for phytoplankton is essentially that of a dilute solution of the chemicals printed on the contents list of a bag of garden fertilizer. Fertilizer added to a lake will usually accelerate the growth of phytoplankton. Light is the energy source for plant growth. Light is abundant near the lake surface. Light is absorbed by the water, so the deep water of lakes is usually too dark to permit photosynthesis.

The light intensity that just permits a photosynthetic organism to produce enough sugars to meet its own metabolic needs is called its compensation point. The compensation points of the different photosynthetic organisms are sufficiently similar that we can refer to a compensation zone. How deep the compensation zone is depends on how bright the sunlight and also on what coloring material or suspended particles are present in the water. Below the compensation zone plants die.

Some lake water is almost tea-colored from dissolved plant material, and some is milky with suspended clay. Also, phytoplankton particles themselves block the passage of light, so that a highly productive lake or pond may have a compensation zone of only thirty centimeters (around twelve inches) below the surface while an almost sterile lake may have a compensation zone tens of meters down.

Nonphotosynthetic organisms in the lake feed directly or indirectly on the photosynthesizers while taking dissolved oxygen out of the water and adding carbon dioxide to the water along with all their other wastes, both liquid and solid. These wastes may be ultimately broken down and are in fact the source of chemicals used for growth by the phytoplankton. Organic debris that does not break down completely rests on the lake bottom as mud.

If the phytoplankton growth rate and all the feeding and decomposition rates were just right, it would be possible to maintain an ongoing cycle of nutrients from water to phytoplankton to herbivores, predators, detritovores, and back to nutrients—a balanced aquarium. A balanced aquarium breaks down if:

- All of the plants or herbivores are eaten out of the system.
- The dissolved oxygen supply runs out, so the animals and bacteria can no longer function.
- The container fills up with garbage into which dissolved oxygen cannot circulate.

While lakes are never balanced aquaria (in fact, balanced aquaria of any level of complexity do not exist), they are convenient images for understanding some of the overall processes that connect all the organisms in a community with each other and with the physical and chemical properties of the environment.

Thoreau speculated on how lakes changed with time. He suggested that:

> The phenomena of the year take place every day in a pond on a small scale. Every morning, generally speaking, the shallow water is being warmed more rapidly than the deep, though it may not be made so warm after all, and every evening it is being cooled more rapidly until the morning. The day is an epitome of the year. The night is the winter, the morning and evening are the spring and fall, and the noon is the summer.

I will describe a year in a typical small lake.

In the temperate zone, the most obvious change during the year is that the temperature changes with the seasons. Starting in the late fall, the pond will be covered with ice. Ice, snow, and water absorb light. By early spring the water under the ice has been protected from the wind for several months. During the winter, the bottom water has probably been too dark for very much photosynthesis. The more superficial water has had adequate light, but nutrients for plant growth have been in short supply because under the ice material could sink but not be restored to the surface water by wind mixing.

When the ice froze on the surface, the surface water was at 0°C and the deep water was at 4°C.* It has had no chance to warm since then. Therefore, when the ice melts the water temperature at the surface will still be zero next to the ice and the deeper water will be at 4°C.

If the water temperature (and therefore density) is the same all the way down, the resistance to mixing is minimal. The wind need only overcome viscosity, the stickiness of water, which is very low. Therefore, immediately after the melting of the surface ice a relatively gentle wind can thoroughly mix the water in the lake.

Deep mixing by the wind after the ice melts is important enough to have its own name: the *spring overturn*. The spring overturn brings nutrient-laden bottom water, which has been lying in darkness under the ice, up to the light, so that phytoplankton near the surface can grow rapidly. The overturn also brings oxygen from the surface water into the lake depths, which aids the metabolism of animals, fungi, and bacteria that may have been starved for oxygen. The spring overturn results in a burst of biological activity called the spring bloom.

As the season progresses, light energy entering the water raises the water temperature starting at the top. The wind dis-

---

* Recall that water expands as it freezes and the expansion starts when it is at 4° C.

tributes this top layer of warm water more deeply into the lake. As the surface water warms, it becomes less dense. With the passage of time stronger winds are required to completely mix the lake. By early summer the mixing process no longer extends into the deeper water. A relatively thick, wind-mixed, warm layer develops on top of the lake, floating on the colder, deeper layer.

The upper layer is called the *epilimnion* (i.e., the upper lake) and the colder, deeper layer is called the *hypolimnion* (i.e., the lower lake).

Between the two layers is a region in which temperature decreases rapidly with depth. This is called the *thermocline,* or region of rapid temperature change. A light wind can easily mix the warm water down to, but not through, the thermocline. It takes very strong wind to mix water through the thermocline. The hypolimnion remains relatively cold and out of direct contact with air or wind.

Once a thermocline has formed, the lake is said to have entered the period of summer stagnation. Life in the epilimnion and the hypolimnion becomes distinctly different. In the epilimnion phytoplankton face two immediate problems and one long-term problem. The immediate problems are, first, being eaten by zooplankton or other consumers, and second, if phytoplankton sink through the thermocline into the deep water, they may enter a region of light shortage, so that their photosynthetic activities can no longer match their own metabolic needs and they die.

Their long-term problem is that at the spring bloom nutrient material will have been incorporated into the body of the producers. If a producer drops into the darkness below the thermocline, the nutrients bound into its body will be added to the hypolimnion. Whenever one organism eats another, crumbs of the meal will sink. Also, there is a rain of excreta sinking into the hypolimnion—loose fish scales, feces, body slime, and all the garbage of the sort that clouds an aquarium. All of these particles carry nutrients that will no longer be available to the phytoplankton in the epilimnion. The pace of life in the epilimnion will therefore decline due to lack of nutrients as summer stagnation goes on.

During the spring bloom there will be a great deal of activity in the deeper water of the lakes. There will be plenty of oxygen dissolved in the water, and all kinds of organisms and organic materials will be swept down to the bottom, where the worms, bacteria, molds, and others will consume them, returning their component chemicals to the water. After stagnation starts there will be very little new oxygen available to the benthos.

The darkness at the bottom is why there are no rooted plants in the middle of reasonably deep lakes and why aquatic plants often have special mechanisms for growing up from the darkness of the bottom toward the lighted surface. Water lilies have energy-rich, nutlike seeds, and the roots of water lilies are rich in starch. They were eaten for their starch by some of the American Indians. In the spring the stems of white water lilies are extremely thin threads that grow rapidly upward until they reach the compensation zone, and then they produce broad, flat leaves that will lie on the surface.

By the end of summer, life in the epilimnion slows down due to lack of nutrients. Life in the hypolimnion slows down, despite an abundance of nutrients, due to a lack of light for the producers and a shortage of oxygen for the benthic consumers.

The cool weather in late summer and fall lowers the temperature of the water at the surface of the epilimnion, so relatively gentle winds can mix that surface water into the deeper epilimnion. Eventually the entire epilimnion approaches the temperature of the thermocline and the thermocline disappears. The next wind can stir the lake deeply, bringing oxygen to the deep water and bringing nutrients from the dark, deep water into the light. This is the *fall overturn*. It permits the beginning of the fall bloom—a new period of high biological activity. This can last until the ice once again covers the surface and prevents the wind from stirring the water.

Under the ice, winter stagnation begins. Nutrients sink to the bottom. The decomposers consume the organic material, releasing mineral nutrients into the water, and the lake waits for spring. The next year it will all be repeated, or will it?

In some lakes all organic material produced in the lake is broken down to mineral salts—the plant nutrients—and carbon dioxide. These are called *oligotrophic,* or low-nutrient, lakes. In these lakes there is enough oxygen to permit all the dead organic material to be digested by bacteria and molds. Oligotrophic lakes are clear and often beautiful, and have clean sand on their bottom.

Most oligotrophic lakes are found in places with very low human populations. You can eat their fish and cautiously drink their water.* Visitors can bathe in the lake, wash their clothes and dishes, dispose of their personal wastes in the water, and probably give their friends a glowing account of their vacation; no harm is done because the lake is starved for nutrients. Organic materials merely provide nourishment for a few more trout.

Any newly formed epilimnion seals the hypolimnion from the air. In the trapped water there will be a quantity of dissolved oxygen, and it is this dissolved oxygen that must supply the needs of the decomposers and animals throughout the period of summer stagnation. If the volume of the hypolimnion is sufficiently large, or if the total amount of organic material that will drop into the hypolimnion is sufficiently small, the organic

---

* Caution is required because of the recent spread of dangerous protozoans from beaver feces.

material can all be decomposed, producing energy, carbon dioxide, and inorganic nutrient salts. The oxygen can be replenished during the fall overturn. In winter the ice prevents the wind from mixing the water. If there is plenty of oxygen, all organic matter can be decomposed before the ice melts in the spring.

If all the organic matter in a lake is decomposed in a timely fashion, there are no organic sediments deposited on the bottom. That is why oligotrophic lakes have clean sandy bottoms. If there is no organic matter left in the water, it also tastes clean. Reservoirs of drinking water are kept oligotrophic if at all possible.

But when too many people use the lake as bathtub, sewer, and resort, a dangerous limit may be approached. If the total volume of the hypolimnion is relatively small compared with the total organic material produced by the lake, the oxygen in the hypolimnion may become exhausted before the decomposers have finished their work. The partially decomposed material will then fall to the bottom and produce mud.

Any lake producing bottom mud is called *eutrophic*. The space taken up by the new bottom mud is subtracted from the hypolimnion volume the following year, making the problem worse. A eutrophic lake is en route to filling up and disappearing as a lake. Reversing the process requires that conditions change.

Lake Tahoe, on the California-Nevada border, was an ideal oligotrophic lake that attracted too many visitors. One year it suddenly developed a muddy bottom and water opaque with plankton. The problem was solved by diverting the flow of sewage produced by the town of Tahoe out of the Lake Tahoe watershed.[25]

Clearly, lakes will differ. The differences depend on many things. Their size makes a difference—small lakes do not contain very much oxygen in their hypolimnion, and very small lakes never form a thermocline at all.

Some lakes have extreme chemical composition, which severely reduces the number of kinds of organisms found in them. The Great Salt Lake of Utah contains just a few microorganisms, beautiful brine shrimp, an abundance of flies, and blue-green algae.

Around the world there are ponds near the seashore that are occasionally flooded with salt water. In some of these rain and runoff place a layer of fresh water above the salt, producing a sharp density difference at an intermediate depth—a *chemocline*.

On the coast of the Sinai peninsula, a dozen kilometers south of the city of Eilat, is found a small pond called the Sun Pond. The salt water lies in the bottom of the pond, and the sunshine easily penetrates the transparent fresh water that overlies it. The sun's energy warms the deep salty water. Because no mixing occurs, swimmers in the pond experience shocking warmth when they insert a foot through the chemocline. There has been cheerful talk of gaining electric power by taking advantage of temperature differences between different layers of water, but so far this has not proved practical.

The local climate makes a difference to lakes. In the tropics the thermocline may never be broken by surface cooling. In the Arctic there may only be a brief ice-free period permitting mixing of the water by the wind, without any thermocline ever forming.

Rivers differ from lakes in many ways. Flood season and dry season strongly affect rivers. There are white and black branches of the Amazon River. The white branches carry a load of silt, which gives color and also supplies mineral nutrients. The black branches, with water the color of tea, can be so free of nutrients that rainwater actually increases the supply of mineral nutrients for the plants.

In much of Amazonia the annual flood of the river extends miles into the forest. Canoes can move beneath the trees. Once, as a visitor at the Smithsonian laboratory near Manaus, I rode in a canoe beneath the trees during the flood. We stopped under a rubber tree, clapped our hands sharply, and were almost immediately surrounded by large, evidently hungry fishes. These fishes were mouth-breeding cichlids that swim into the forest and search for nuts and seeds. The clapping imitated the sound of rubber seeds popping from their pods. During the flood the fish grow fat, and as the water recedes they return to the riverbed, where there is no food for them.

When these fish breed the male takes the fertilized eggs into his mouth, where they grow on their own yolk and mature to a swimming stage. This may take weeks. During these weeks the male cannot feed, but there is nothing much to feed upon until the fishes can once again leave the riverbed for the fruit and nut harvest. Brooding young may be the best thing for the fish to do with its mouth under the circumstances. The fascinating story of the fish in the forest illustrates that sometimes very particular circumstances can define the properties of a community.[26]

We have seen that there is a basic and recognizable pattern of events that holds in a general way for all lakes. We have also seen that chemical, physical, climatological, and geological peculiarities are modifiers of this basic pattern. I suggest that it is possible to gain a kind of understanding of all communities on earth by considering them all as more or less extreme variants of lakes. Essentially the same functional groups exist in any natural community. Organisms participating in each one of these lifestyles have certain generic resemblances, regardless of their body size or where they are found. Herbivores, carnivores, detritovores, and parasites are everywhere, differing from place to place in their details.

Often members of a large evolutionary family may be spread around the world and have approximately the same role everywhere. For example, predacious cats cover the globe. Lions are big carnivores in Africa, tigers in India, pumas in the American Southwest, lynx in the north-central plains of North America, and jaguars in Central America.

This is simplified—there are other cats and other places—but my point here is that functions persist over space and that sometimes slightly different species are arranged in an approximately mutually exclusive geographic network. Some cats chase down their prey; others lie in wait until the prey are close enough and then attack. While there are many kinds of cats found in different landscapes, none of them have become herbivores.

Cats are elegant predators but certainly not the only ones. Everywhere on dry land there are other large and small predators, pouncing or stalking or lying in wait. They include birds, reptiles, scorpions, insects, and more.

Returning to the water, most fishes are carnivores. Beetle and dragonfly larvae stalk the waters of lakes and ponds, attacking zooplankton. Polyps and jellyfish cannot swim very rapidly but wait until their prey comes to them.

Similar remarks apply to herbivores. How to be herbivorous depends on location and circumstances. The important herbivores may be anything from bison to clams to caterpillars to copepods.

Many animals operate on several trophic layers at once. It is particularly common to find carnivorous predators moonlighting as detritovores. Examples range from flatworms a quarter of an inch long to 150-pound hyenas. Even early humans and their ancestors weren't above eating the leftovers of other hunters. The fascinating details have filled and will continue to fill many ecologists' lives.

Organic material in lakes and rivers is at least partially decomposed by bacteria, and nutrients are removed from the water by plants. Water from which all the organic material has been removed is suitable for drinking, even though it may have a fair quantity and assortment of minerals. The minerals lend flavor and unless they are very concentrated are usually not harmful. Naturally clean water is relatively free of organic material and bacteria and has almost no taste. Only distilled water is free of minerals, but it has a flat taste. Usually attempts to chemically lower the bacteria count of drinking water cause unpleasant tastes. I find chlorinated water to be unpleasant. Connoisseurs consider bourbon best when drunk with "branch water," unpurified water from a natural stream.

For thousands of years, pipes and aqueducts have brought clean drinking water great distances to flow into pipes or wells. Because water passing through human settlements is changed, water entering settlements is always first used for drinking.* Drinking tap water was standard in most of America in the mid-twentieth century. In the last half century water quality was thought to be deteriorating in many American cities, although actual evidence for seriously deleterious effects of most tap water is not very strong. By the century's end drinking bottled water, sometimes from uncertain sources and at the price of cheap wine, became very common.

---

* In the town of Bethlehem there still exists the water works of King Herod the Great. He arranged to use the water supply for many different purposes. At the spring some water was diverted into a small open aqueduct, around fourteen inches wide. This, the purest water, flowed by gravity all the way to the city of Jerusalem. Excess water was drained into a pool that was used for local drinking water. The overflow from this went to a pool for swimming, surrounded by benches for sunbathing. This pool was reached by steps. The next pool was reached by ramps suitable for herds and was used for animals. The overflow could then be used by agriculture. This elegant system was built for a small town two thousand years ago.

I have introduced the point that approximately the same processes of eating and being eaten, escaping and not escaping, reproducing and dying occur in all of nature everywhere, with different actors and different details. Everywhere there is a release of nutrients into water by decomposition and an uptake of nutrients from water as plants grow. There is always dependence on free water.

Oceans

The world's oceans may be thought of as extremely large lakes, with critical differences.

There is a compensation region in an ocean, just as there is in deep lakes. Recall that this is the depth at which light is so reduced that photosynthesizers can produce only enough to maintain their own respiratory needs, and obviously cannot support herbivores. The compensation point may be as deep as three hundred feet in some of the clearest parts of the ocean or as little as three feet in richly productive waters near the shores.

Oceans receive fresh water from rain and runoff and rivers, but they are salty. Where did the salt come from?

Over billions of years rainwater and river water have drained into the oceans and ultimately evaporated to fall as rain once more—an endless cycle of water. The early oceans were much less salty. If we boil water in a teakettle, a visible crust of minerals soon forms inside the kettle. The boiling water that evaporates leaves its dissolved minerals behind. As water evaporates from the sea it also leaves its minerals behind, so that the salt content of the oceans has been gradually increasing for the last four and a half billion years.

The oceans can also lose salt when arms of the sea are cut off and the water in the arm evaporates, forming solid salt deposits. Some of these deposits are so large that they can accommodate systems of roads, such as those underlying the city of Detroit,

Michigan. The Wieliczka salt mine near Krakow, Poland, contains a large, almost cathedral-like church carved completely from salt. Nevertheless, losses of salt from the sea are very small compared to the sea's total salt content.

The salt concentration in the sea varies from place to place depending on rainfall and evaporation patterns, land drainage, and the history of the water mass. By contrast, the surface waters of most lakes are not chemically different from spot to spot within a lake.

The effect of rotation of the earth is trivial in lakes, but the size of the oceans means that it has an important effect on ocean currents. This is called the *Coriolis force* and requires some explanation.

The earth revolves 360° each day. Although the daily rotational velocity is 360° at both the equator and the poles, linear distance moved during the day because of the earth's rotation varies with latitude. At the poles the distance is zero. At the equator the distance traveled is equal to the circumference of the earth, about twenty-five thousand miles.

Imagine a traveler moving across latitudes. The latitudinal difference in velocity sets up an apparent acceleration to the right of a moving object in the Northern Hemisphere and to the left in the Southern Hemisphere.

I find I can visualize how that works by imagining myself crawling on a rotating phonograph disk. Each point on the disk has the same angular velocity, but the linear velocity is much greater near the disk's circumference. If I try to move in a straight line across the disk, the section of my body further away from the center will be moving faster than the other end.

If I am small enough, the difference in movement between my front and back ends doesn't matter. If, however, I think of myself as a caterpillar crawling on the same disk, one end of my body will be fairly strongly shifted compared to the other. I would have to exert force to overcome the effect of rotation. That force I exert is called the Coriolis force.* It matters very much for ocean surface currents, turning surface currents to the left in the Southern Hemisphere and to the right in the Northern Hemisphere.

The circulation pattern of the oceans is driven by the major surface winds and by the fact that the water's saltiness and temperature together determine its density. While each region of the ocean has characteristic temperature, salinity, and biological populations, the actual water is changing all the time. The oceans must be thought of as great tubes, thick sheets, and streams of water moving over, under, or past each other, mixing in some places but not rapidly enough to obliterate the ropy structure. Each water mass is moving at its own rate, which is a function of wind patterns and its own past history and of how it is related to the surrounding water masses.

Both temperature and salinity have a significant effect on water density. Where two large water masses meet, the denser tends to slide under the less dense, which may mean the saltier under the less salty or the colder under the warmer.

This gigantic writhing liquid mass is on a spinning globe, and so the Coriolis force tends to pull the earth out from under the ocean to the east, or conversely gives the water currents a tendency to move toward the west as seen by someone standing on our globe. There are also continents sprinkled over the globe. The upshot of all this is that near Antarctica—south of the other continents—the surface currents of the ocean move completely around the world from east to west. Further north, the surface flow is interrupted by continents.

---

* It is not a proper physical force acting on me. I know it is there only if I insist on walking in my chosen direction, despite the rotation of the floor beneath me.

At the equator, in both the Atlantic and the Pacific, there is a westward-flowing current driven by the trade winds. The westward flow is interrupted by South America in the Atlantic and by Asia and Australia in the Pacific. When the equatorial surface currents reach the continents they divide into branches flowing to higher latitudes. These warm streams—the Gulf Stream in the Atlantic and the Japan Current in the Pacific—bring warm salty water to regions of cold and rain, where part of the water sinks into the depths and proceeds toward the Antarctic.

The net effect of all these restrictions and motions is that on the surface of both the Atlantic and Pacific Oceans there are two gigantic swirls of water. Above the equator the water moves in a great clockwise swirl in each ocean and below the equator in a great counterclockwise swirl. For example, in the Pacific Ocean water near South America there is a surface current near the equator that moves water from the coast of South America toward Asia. This circulation pattern is of enormous importance for life on earth in many ways.

These water masses are so large that small organisms—phytoplankton, bacteria, zooplankton—live their lives within a single water mass. As the water mass moves westward the same kinds of events transpire as in the epilimnion of a lake. However, what occurs in a time sequence in a lake occurs in a spatial pattern in the ocean. Phytoplankton is eaten. Zooplankton may die. Fish feed on zooplankton and on each other. The dead and bits of organic garbage sink out of the floating water mass into the dark depths, carrying nutrient material with them. The further off the coast, the less fertile the water—just as the longer the time into stagnation the less fertile the epilimnion.

The surface layers of the Pacific Ocean lose nutrient material as they travel along the equator from South America to Asia; in the Atlantic this happens as the surface layers move from Africa to the eastern coast of South America.

The average depth of the oceans is more than two miles—more than ten thousand feet. Only the top skin, less than three hundred feet thick, produces almost all the organic material. In the oceans, as in lakes, the main producers are phytoplankton. There are some floating larger seaweeds, such as *Sargassum*. Some attached vegetation, such as the giant kelp near the California coast, grows hundreds of feet upward toward the light. These giant kelp forests and floating masses of seaweed are not of major global significance, but they do shelter interesting and important species such as sea otters, which would probably become extinct if these plants were gone.

At some volcanic hot springs that open onto the sea bottom it has recently been found that bacteria can produce organic compounds in darkness, capable of feeding a broad array of animals including large wormlike things, large clams and crustacea, and peculiar kinds of fishes. The bacteria get energy from inorganic chemical reactions. The total amount of material produced is small on the scale I am considering, but it is a fascinating phenomenon that deserves mention.[27]

Large, deep lakes may not turn over on any regular time basis. For example, Lake Tanganyika usually has a definite chemocline, with oxygen-poor, nutrient-rich water lying below the relatively sterile epilimnion. Occasionally the water in the lake can be set swinging by an appropriate sequence of winds, like a plate of soup carried by an inexperienced waiter. This can splash nutrient-rich water into the surface layers, helping to maintain the fisheries of the lake.

But oceans never turn over.* As in a lake, producers that sink below their compensation point for any considerable length of

---

* There is a possibility that the entire Mediterranean Sea did turn over, like a gigantic lake, once or twice during the Pleistocene. The low-oxygen bottom water was a major mortality source for the shallow-water organisms before it had a chance to equilibrate with the atmosphere.

time die, and their remains may continue sinking into darkness. Since oceans never turn over, there is no return from the depths for oceanic phytoplankton.

Organisms of the same specific gravity as the surrounding water do not sink. The presence of a hard shell may increase specific gravity, but many ocean phytoplankton and zooplankton have devices that slow their rate of sinking. One common property is the presence of oil droplets inside the organisms. In this case, fat is not necessarily only a way of storing energy but also a flotation device. The other general antisinking mechanism, used by both plants and animals, is the growth of amazingly elongated body extensions—beautiful plumelike excrescences—which slow the sinking rate by enhancing the viscous drag of the water.

If oceans never turn over, how can the near-surface phytoplankton get enough mineral nutrients to maintain themselves? In much of the ocean they cannot. The blue-water regions of the travel posters, like the clear waters of oligotrophic lakes, are largely biological deserts. On the other hand, the fertility of the eastern Pacific on the coast of Ecuador, Peru, and Chile during a "normal" (or La Niña) year is enormous. Until the last two decades the fish yield from these waters was one of the highest in the world, and even now examination of a can of tuna often reveals a South American origin.

The fertilization occurs because, as the coastal current moves north, winds and the Coriolis force peel the surface water away from the coast, causing an upwelling into the light of deep water, some of which has been accumulating nutrients for centuries. Nutrients that had been unavailable to photosynthetic organisms are locally abundant in regions of strong upwelling. Notice that the enrichment is spatially rather than temporally limited. Where the biological activity is great enough the rain of organic matter to the bottom may be too great for the local dissolved oxygen supply, so decomposition stops with a layer of bottom mud, like that in a eutrophic lake.

The biological activity in upwelling regions is enormous. I have been told that in the upwelling regions of West Africa, the Soviet Union built fish-processing plants on the assumption that trawlers could consistently fill their holds with fish in a matter of hours.*

By contrast, when the wind direction reverses in upwelling regions, such as the west coast of South America, the biological activity slows down dramatically. These are the El Niño years. I will not discuss them here. The conditions for El Niño or La Niña hinge on global climatological issues beyond any direct consequences of ecological changes, although their ecological effects are enormous.

The effect of large rivers is dramatic. The outflow of the Mississippi River produces a broad tongue of fertile water more than a hundred miles long and fifty miles wide; traces of the Mississippi outflow can be followed much farther than that.

Occasionally bubbles of seawater mixed with river water maintain their integrity for many days. These bubbles carry their own organisms with them, which occasionally produces relatively rich plankton populations in the open sea while the surrounding water remains blue and sterile. I found a bloom of a colored ciliate (*Mesodinium pulex*) in a tongue of dilute water ninety miles south of the mouth of the Mississippi River on June 2, 1952.[28] The same kind of regional fertilization is found at the mouth of all major rivers.

Upwelling regions, along with estuaries and regions that receive drainage from the continents, are the source of more than 90 percent of the world's supply of edible fish. When the water leaves the continent and flows out into the ocean it enters a con-

---

* Personal communication from the late Howard Sanders.

dition similar to that of summer stagnation in lakes, in which light is present but nutrients are diminishing, to be replenished from deep, dark waters in upwelling regions. The time between sinking and upwelling for any single water molecule may be thousands of years.

Larger animals can swim through the boundaries between water masses. Whales, large fish, and sharks roam the oceans at will. Their will is often influenced by hunger. For example, the enormous krill-eating blue whales feed along the borders between large water masses. These borders tend to have higher productivity of phytoplankton and therefore of krill. The higher productivity seems due to the fact that a particular collection of phytoplankton growing in one water mass is likely to exhaust its supply of mineral nutrients. Another water mass, having a different history and a different set of phytoplankters, may exhaust the nutrients needed by its phytoplankton, but there is the possibility that the precise pattern of exhaustion will not be quite the same. For example, one water mass may lack nitrogen and the adjacent water mass may lack phosphorus. Where the two water masses meet and some mixing occurs there will be a richer food supply than in either water mass separately.*

Obviously there must be some way by which fertilizer is returned to the ocean's surface water, or life in surface waters could not continue. The fertilizer brought to the sea by rivers is derived from the rocks that were once parts of the sea bottom, as can be easily demonstrated by the presence of fossils of recognizable sea animals in them. They have risen from the sea during the past billion years.

There is a slow cycle that moves nutrients from the very deep water of the ocean. The transition of an atom of phosphorus

---

* Personal communication from Phillip Dustan of the University of South Carolina.

from sea surface to sea bottom to sediment to underwater rock to uplifted rocky hills to solution in rivers and then back to the sea may take from tens to hundreds of millions of years.

There is also a much quicker cycle occurring, in which a circuit of a fertilizer atom may take only thousands of years or even less. This is the process of upwelling, which sets the location of the most important marine fisheries. Upwelling occurs wherever a current is flowing toward the equator along a steep continental slope on the west coast of a continent. It also occurs when currents flow away from the equator on the eastern slope of continents.

Dry Land

Most general ecology books spend most of their pages on terrestrial situations, although dry land covers less than 25 percent of the earth's surface. Water and ice cover the rest. Dry land carries an enormous diversity of species and of kinds of associations among species.

We are terrestrial organisms. Most people's crops, dwelling places, and commercial activities are on dry land.* *Dry land* means land not actually under free-flowing water. I will now use our earlier focus on wet ecosystems to simplify our discussion of terrestrial ecology.

Dry land ranges from damp meadows and forests to sandy deserts. Really dry land, of course, cannot support life at all. Nor can damp sand. What is required is soil.

Soils are enormously different from place to place. There are poor soils, such as the spoonfuls of grainy material that support plant growth in the chinks of rocks, to rich soils, such as those found in temperate-zone hardwood forests. These are mixtures

---

* People in Venice, people in many villages in Papua New Guinea and Bangladesh, and various marsh and boat people are notable exceptions.

of sand grains, water, rootlets and broken bits of leaves and stems of dead plants, tiny animals and their droppings and molts, fungal hyphae, bacteria, protozoa, and the greasy, partially decomposed organic mud that stains hands and clothing. Rich soils also contain vitamins and trace minerals.

The various bacteria in soils each consume some organic compounds and secrete others, resulting in an elaborate exchange of chemical materials, ultimately acting to mineralize organic compounds.

Not everything buried in the ground decomposes, and not all soils are equally effective places for decomposition. (Recall the speech of the grave digger in *Hamlet*.) Nevertheless, organic material enters the soil and dissolved mineral salts leave it, either as drainage or by entering plant roots. Nutrients are regenerated in the soil and are dissolved in water as they are in the benthos of a lake.

Different landscapes have been named by a variety of systems. Sometimes areas are named for their most conspicuous vegetation. The shape of the vegetation divides landscape into forests, meadows, grasslands, savannas, steppes, marshes, dry deserts, rain forests, and also croplands and other areas directly modified by human activity.

Each area can be a focus in its own right for a lifetime of study. Each contains producers, predators, symbionts, mimics, parasites, and hyperparasites. Even the hedgerows that line country roads, particularly in Great Britain and France, have merited careful study by ecologists.

Named pieces of landscape are sometimes thought of as "communities." It is assumed that interactions between organisms are more intimate within communities. Often maps are drawn showing which communities occupy which areas. The borders between areas are considered important as indicators of changes in soils or other environmental conditions.

But even the border between dry-land communities and aquatic communities is not always obvious. In southern Florida there are mangrove reefs where the land meets the Gulf of Mexico. They consist of open forests of mangrove trees growing in water. The mangrove roots house an assortment of sponges, fishes, and other animals, with open-water passageways separating individual trees. The richness of organisms is reminiscent of a coral reef or tropical forest.

In central Florida are the Everglades, consisting of grasses and shrubs living in a very broad shallow river during the wet season, interspersed with hammocks bearing tall shrubs and trees. In the dry season the flat Everglades are subject to burning, and free water is found primarily in pools dug by alligators.

There is a line on maps of Florida to separate mangrove communities from the Everglades. The line is a simplification. Driving east on the Flamingo Highway demonstrates that at the very edge of the Gulf of Mexico there are only mangroves. Further east occasional patches of Everglades grasses appear between mangroves, and as one continues, the grassy patches become larger and then coalesce so that there are patches of mangroves surrounded by grass. The border is fractal-like—it cannot be made to appear smooth by altering the scale of observation.

In the Wabash Valley of Indiana are found maple forests and forests of evergreens. The border between these two well-defined and carefully studied types of forest is also fractal-like.[29]

All terrestrial organisms rely on the convenient presence of water, so that rains, snows, dews, rivers, springs, and lakes are vital.* In describing any kind of terrestrial community the water relationship will be the first item of the description.

---

* Some desert animals do not drink but get their water from metabolizing carbohydrates.

The later parts of this book will tend to focus on the terrestrial aspects of ecology because terrestrial problems engage popular concern. However, I will not examine the full panorama of terrestrial communities. I will give a rough description of one example of a terrestrial system and will assume that all the others function in approximately the same way and that the way they function at a general level is essentially the same as the function of watery systems.

For convenience, I will choose a hypothetical mixed hardwood second-growth forest on Long Island and follow it through a year, as I followed a hypothetical lake.

Start at a time in the fall when most of the trees and bushes drop all their leaves and many of their twigs. The annual herbs are dying and lie withered on the ground. Grass is turning brown. What happens to dead plants and parts of plants when their season has passed?

The dead plant material forms a detritus layer, analogous to detritus from the productive zones of lakes or oceans. If water is abundant, decomposition of dead leaves by bacteria and fungi starts immediately. In many slightly drier woodlands the ground will have a top layer of dry brown leaves that are not decomposing and which act as a roof, preventing excessive drying of the layers underneath. These dry leaves may wait a year until new dead leaves cover them and they become damp enough to decompose. By late fall a Long Island woodlot will have around four layers of leaves on the ground.

The first signs of decomposition may be white threads on the underside of the leaves and twigs. There will also be a characteristic odor coming from the damp leaves, similar to, but pleasanter than, that of crushed mushrooms or an unwashed refrigerator.

Insects, mainly beetles, caterpillars, and sap-sucking bugs, will have minced some leaves and drained the sap of others before they fall from the trees. If there is lime in the soil, there may be earthworms pulling leaves down into their holes. Beetles and pill bugs will chew at dead leaf fragments. Most of the organisms that contribute to the destruction of dead leaves are much smaller and stranger than these.

In fact, the total number of animal species turning fallen vegetation into soil is often measured in hundreds even in a one-hundred-square-centimeter area, without counting protozoa and bacteria. They are the detritovores and their predators—snails and slugs, myriads of little millipedes, beetles, mites, threadworms, and even terrestrial crustacea such as pill bugs and sowbugs. There will also be collembolans—tiny wingless insects with a spring mechanism on their abdomen that will throw them many body lengths into the air when they are disturbed. These are among the commonest of arthropods, but except by some biologists they are among the least frequently seen.

These detritovores will all be chewing the dead leaves into bits, making them more accessible to the bacteria and molds or eating the molds and bacteria that are already on the leaf fragments, thereby exposing the fragments to new bacterial settlement.

There are also tiny predators in the soil. These include predacious mites, small centipedes, predacious nematodes, and insects. My favorites are tiny pseudoscorpions—smaller than sand grains—which closely resemble their more dramatic large relatives but lack the stinging tail.

There are amoebae insinuating themselves between the damp particles, eating bacteria. Some of these, the slime molds, are sometimes individual amoebae, but they can aggregate and form yellow or orange sheets or ribbons that slide over surfaces, consuming bacteria. When the local bacteria supply runs out or the physical conditions change, these organisms can aggregate to form themselves into vertical columns topped by globes composed of spores that can be blown large distances by the wind, landing to hatch as single amoebae in any wet, bacteria-rich landing spot and start the process again. It is not clear if they should be considered fungi or protozoa.

This remarkable menagerie of tiny animals can be seen if you have access to a relatively low-power microscope—enlarging perhaps fifty or a hundred times. Some preparation is needed since they are hiding between the mineral grains and the organic debris that makes up most of the soil.

Place a tablespoonful of surface soil and debris from a meadow or a woodland floor in a small tea strainer or any container with a fine screen bottom. Place the container over a pool of clear alcohol (vodka will do). Then arrange an electric light bulb several inches over the soil-filled container. The bulb dries the top of the soil, the microfauna burrow deeper, and eventually many tiny animals fall through the mesh into the alcohol, where they are preserved for your viewing pleasure.

If it all works well, it is as fascinating as a jungle safari, but there are several reasons it may not work. Excessively dry soil or excessively wet muck may not contain any active organisms, the first because the organisms dry up or leave, the second because there is no available oxygen.

If the soil is from a lawn that has been subjected to recent poisoning against weeds, moles, or insects, the normal soil animals may all have been killed. A newly plowed field or newly cultivated garden soil may have been so disturbed in its granular structure that the animals have left or died.

Surveys of soil microfauna that actually attempt to enumerate the number of kinds of organisms in the samples are enormously labor-intensive, and not many are undertaken, although the general method of extracting the organisms has been known for a hundred years.

In a study of the microfauna of an old field in Michigan surrounded by forest, the forest soil organisms were very different from those of field soil, but a single tree out in the field cast a shadow of forestlike organisms between itself and the forest.[30] Why this occurred is unknown.

The tiny soil-dwelling organisms are still so unstudied that any shovelful of dirt is likely to contain two or three new species.[31] I do not know, and I don't think it is known, how widely or narrowly species of forest soil organisms are distributed. Is there a chance of wiping out a species if we plow a field or weedproof a lawn? Perhaps. Is there danger in this? Probably not. How do I know? I'm guessing.

There are organisms that become intimately connected to the rootlets of plants. Some of these, the mycorrhizal fungi, draw sugars from the plants, but in exchange their threadlike bodies act as extensions of the root system, providing water and minerals to the plants.[32] These fungi may also form a filamentous network connecting fairly distant plants to each other. This permits nutrient exchange between trees but also allows the spread of viral and other diseases.

Parasitism of animals by molds, bacteria, protozoans, and other animals, and parasitism of plants by animals, of bacteria by viruses, and so on, is rampant in the soil and in all terrestrial and freshwater situations.[33]

Protecting plants from parasites occasionally has unpleasant side effects. For years extremely toxic chemicals were added to the soil on eastern Long Island to protect potato plants from the golden eel worm, a little nematode that infects their roots. This poisoned local wells. The present solution is to drink bottled water and to switch from growing potatoes to growing grapes, which are afflicted with an entirely different selection of pests and parasites, requiring quite different poisons to control them.

Soil ecology is subject to fairly extreme changes. Some of these are short-term—too much or too little rain, a freezing night—but the overall process of decomposition and release of nutrients is unlikely to be interrupted except by events that actually kill off most kinds of soil organisms. If one or two species vanish locally, it probably doesn't matter.

If all is well, the energy in the detritus is being used by the molds, bacteria, and animals while the minerals are in the process of being released from the detritus, passing through the detritovores and becoming soluble in water. The nutrients may then be drawn into the roots of plants or perhaps washed away to a stream and ultimately to a lake or ocean. In the process carbon dioxide is produced, some dissolving in water before it ultimately reaches the air. This wonderfully complex world of soil organisms is vital to maintaining soil quality.[34]

I don't know how long soils persist. Thousands of years is a possibility, in the absence of severe flood or erosion. Fairly often soils are replaced by silty floodplains, windborne layers of powdery loess, volcanic rocks, and dried lake or sea bottoms. Whether or not these will form soil, and what kind of soil it will be, depends on circumstances. Once there is some capacity to store water in or on these mineral aggregations, the process of soil formation can begin, although the busy world of molds and tiny animals usually found in a mature soil may be absent.

Without the chewers of leaves the action of the molds and bacteria is slower. At an extreme, some kinds of dead leaves are peculiarly difficult to decompose. Oak leaves may take several years. The needles of evergreens may be even slower, so a thick layer of dead material piles around the trunks of the trees.

In the northern United States, Canada, and Alaska the season of liquid water may be too short to permit decomposition of the thick coat of fallen conifer needles. Sometimes these become so abundant that they fuel wildfires. Wildfires reduce the organic material to carbon dioxide and ash, which contains most of the mineral nutrients. Unfortunately, ash is easily washed away after a fire.

Algae, descendants of the ancient symbiosis between colored and uncolored bacteria, have difficulty on bare rock because they can be killed by drying, but some cyanophytes, blue-green bacteria, can survive almost complete drying and return to photosynthetic activity when water returns. The cyanophytes are very common. They are very similar to organisms found as the earliest fossils on earth. They can even be found on bare rocks that are splashed by the spray at the edge of the sea. When they are wet they become slightly slimy, but when they are dry these blue-greens look like smears of black paint or tar.

Many rock surfaces, including stone walls and gravestones, are covered with lichens. Lichens consist of algae growing in between tangled filaments of fungi. The fungi provide a store of water while the algae furnish photosynthetic products. Rains permit tiny amounts of mineral fertilizer to leach out of the rock.

Some lichens are thin and flat, but some, particularly in cold places where the availability of liquid water is primarily limited by freezing, can be remarkably plantlike—around eight inches high and resembling dense turf. Reindeer moss is a kind of lichen grazed heavily by Arctic mammals.

Up to this point gravity has been of minor importance in our discussion because organisms are essentially weightless when floating in water. The effect of gravity in water is just sufficiently strong to be used as a direction marker but not strong enough to be a major constraint.

Many organisms living in water have gravity-sensing organs consisting of a tiny crystal or stone in a capsule lined with sensory hairs. Which hairs are stimulated by the effect of gravity on the stone will provide information for orientation. The animals can tell when they are upside-down or right side up.

The largest living animals (blue whale and giant squid) and plants (giant kelp) live in water.* To live out of water requires anatomical and physiological responses to gravity. Placing water animals and plants on dry land usually results in their collapse into a nonfunctional state before dehydration occurs, because of their inability to survive the weight of their bodies. Beached whales die by asphyxiation. Their body bulk out of water imposes an excessive stress on their breathing. Small seaweeds can be taken from the ocean and floated carefully onto a sheet of blotting paper, where they will flatten out to resemble a delicate drawing. Even heavily built seaweeds, such as giant kelp, lie flat on the beach where the waves have thrown them, interfering with volleyball games in Santa Barbara.

If gravity is such a difficult problem, how are we and so many other organisms able to counteract its effects? The same principles apply as used in the engineering of large structures—minimizing weight while maximizing strength. The details are beautiful, and I'm not going to present them. The general effect of scale on the shape of organisms under the influence of gravity has been elegantly studied since the time of Galileo and has resulted in one of the great classical works of biological and mathematical literature, Sir D'Arcy Wentworth Thompson's *On Growth and Form*.[35]

Gravity breaks our bones and destroys our youthful face and figure. The heart pumps blood against gravity. Since the burden

---

* The largest living organisms in one sense are fungi that may infest many square miles of forest soil, or certain kinds of trees if you insist on counting a clone of trees as all one organism, but these are in a sense counted as single organisms by courtesy.

of gravity is proportional to the cube of our linear dimensions, gravity limits the size of all terrestrial organisms. It not only makes obesity dangerous and tree limbs break off but seems to have acted as a constraint on possible evolutionary experiments.

We know from fossils that the first animals on land crawled on their bellies. Among their descendants were some that lengthened their legs and stiffened their spine so that their bellies were no longer dragging. There were others that swung their bodies upright, such as people, dinosaurs, or ostriches and kiwis and cassowaries. Some even defied gravity by flying, thereby finding themselves in a three-dimensional world whose depths we can realize only in airplanes or gliders.

We cannot claim that defying gravity is necessarily a good thing in some absolute sense because there are other descendants of these first belly-dragging pioneers that continue to drag their bellies. Others have lost their legs entirely; some have returned to water, avoiding the problem completely. Those that have conceded the battle to gravity may be failures as acrobats, but they also are evolutionarily successful.

Plants have lifted themselves from the ground to stand as trees, braced against the stress of the wind by their heavy roots and thick trunks. Other plants dispense with the heavy trunks and are vines, holding on to the trees.

Plants must also confront gravity in their production of seeds. A parental plant might "want" a large seed that contained enough stored energy to see the young plant through its early life, but a large seed cannot drop far enough from its mother to avoid her shade. Many different means of seed transport are used.

The seeds of many mangrove trees are large and heavy and are dropped into water and float away from their mother before settling in the bottom. Others are weighted so that they sink into mud before being carried too far. Coconuts, perhaps the largest of all seeds, may float for long distances, germinating hundreds of miles away from their mothers.

Some plants have stayed small and use small seeds equipped with mechanisms for transport by the wind, such as the down of thistles or dandelions. Others have animals carry their seeds—in their guts, like cherries, or on their pelts, like beggar's ticks.

All these mechanisms accomplish the same thing, occasionally bringing a reasonably sized seed to a reasonable place for germination and growth, at least often enough to replace each parent plant once during its lifetime.

Considering that some trees live for many centuries and that, on the average, each adult tree will be replaced by just one adult tree one generation later, it is clear that survival of any one seed is very improbable.

If you try thinking as a plant, you can imagine tricks that might be used to help with growth and reproduction. If your ideas are at all practical, you may be assured that some plants have tried them. For example, there are orchids that mimic the appearance and odor of female wasps so that the male wasps, in attempting to mate with the flowers, pick up pollen and carry it to the next flower.[36]

The invasion of plants onto the land faced different problems from those of animals. On dry land, water is not usually available in the same places where light is available. Plants use their roots to collect water in the dark soil and their leaves to process energy in the light. The plants stretch themselves from one of these locations to the other by using their woody stems.

Again notice the similarity to a lake. The roots are in darkness, where mineral nutrients are available. The leaves are in a well-lit but nutrient- and water-poor region. The trunk connects the two so that growth can proceed.

During the frozen winter liquid water is not available and

growth stops. It is a kind of stagnation. In the spring much of last year's vegetation will have rotted into fertilizer. Spring is a period of rich growth. The water in the ground is once again liquid, and light can reach through the bare stems to the ground.

Because plants on land grow on a hard surface and must remain in a single location, there is a danger of being shaded by competing plants. In many forests it is too dark for photosynthesis. To avoid being shaded out in the spring it may be possible to produce your leaves before the other plants do. Unfortunately, whatever it is that causes the other plants to not be able to grow leaves that early is likely to affect you also.

There are different paths around these problems. One alternative is to become taller than adjacent plants. Another is to be small but fast. A third is to store last year's energy in a bulb or tuber to permit very rapid growth in springtime. Onion bulbs and potatoes give their plants a strong head start. Also, by sharing tubers and bulbs with people, the onions and potatoes ensure that the rest of their offspring will be well cared for, fertilized, and weeded.

My first opportunity to really look at forest floors came when I was a student at a tiny college in West Virginia. There was no botanist on the faculty. I was permitted to teach myself botany by collecting and classifying plants. I was required only to know the names of plants, but incidentally I saw something of how they lived.

In West Virginia most of the trees are bare in winter. As soon as the ground thaws, the little spring-flowering plants—the spring beauties, lilies of the valley, violets, chickweeds, and trout lilies—appear. These are often less than six inches high and cover the forest floor. Many sprout from small bulbs or tubers no bigger than a peanut, which give them enough of a head start to produce seed before they are covered with shade by other plants.[37] A few weeks later coarser plants appear. These include bloodroot, Solomon's seal, Dutchman's breeches, and wild ginger.

Around a month after the first spring beauty, the leaves of the trees and shrubs are becoming full and there are almost no more blossoms on the forest floor except for those that need insects, such as the dark brown flowers of wild ginger, which are pollinated by beetles, and the dark red flowers of the painted trillium, which smell like rotting meat and are pollinated by flies.

By summer the blossoms are past, and often water is scarce and the temperature is uncomfortably high, so plants must spend water to keep cool. In the fall the liquid water disappears entirely and winter stagnation begins.

The ground in many places is fairly well covered with plants, and there is a premium on attaining access to light, despite the shade of other plants. The evolution of tall wooden tree trunks seems an eminently reasonable and perhaps unique solution. They are so big that they cannot be shaded by other plants.

Why don't other trees grow even taller and shade the big trees? Tall trees have a serious problem with wind, and perhaps growing too tall is more of a disadvantage from wind breakage than it is an advantage from the standpoint of shading. But these are evolutionary speculations, which can become dangerous. The size and shape of plants may set up other problems concerning pollination, seed dispersal, wind resistance, and so on. There is probably no unique best solution to problems of evolutionary success.

The tall trunks problem can perhaps be avoided by growing at the edge of the forest or in places where large trees have just fallen down, leaving an open view of the sky. It is also possible to grow tall rapidly without growing a stout stem, such as a vine clinging to some large tree.

If rains are sufficiently frequent, it may be possible to grow high up on some other plant and forgo soil roots entirely. Spanish moss, orchids, and some bromeliads (members of the pineapple family) do this.

HOW INDEPENDENT ARE ECOLOGICAL SYSTEMS?

In the nineteenth century Thoreau and Forbes focused attention on the self-regulating properties of lakes.[38] This focus was strengthened in the mid-twentieth century by a general concern for self-regulating feedback systems.[39]

Lakes were then thought of as sufficiently isolated from the surrounding world so that they were, to a large degree, feedback systems, reusing most mineral resources and controlling population size and species number by internal forces. This view appeared in many elementary texts.

Although I have presented a generalized classical image of an annual cycle in a temperate-zone lake as a way of introducing the idea of ecological communities, current work suggests that the importance of self-regulation in lakes may have been strongly overstated.

Briefly recall the discussion of a small lake: During the winter and summer stagnation detritus sinks from the surface water, and nutrients accumulate in deep water. At the same time, the bottom water loses free oxygen because of respiration by detritovores and bacteria. The photic zone, although rich in oxygen, is biologically quiet because its nutrients have sunk into the deep water.

At the spring and fall overturns, a relatively slight wind can mix the water completely. Oxygen is mixed into the bottom water, so the bottom organisms can continue to decompose organic material, and nutrients mix into the photic zone, so phytoplankton can grow.

This leads to the conclusion that most of the nutrients cycle, from solution to phytoplankton to zooplankton to fishes to decomposers and back to solution. The step from decomposers to solution can occur at any time but may not be as complete or as rapid during the summer and winter stagnation periods. It is not necessarily the case that everything is recycled. If the decomposition process is not complete, mud forms.

Underemphasized in this picture is that lakes take in a great deal of material from their watershed—leaves, pieces of wood, and a miscellany of organic garbage. This exogenous material may be eaten by bacteria, which are in turn prey for microzooplankton, which feed larger zooplankton, which in turn feed fish.

In fact there are two more or less distinct food webs in each lake. One is based on bacteria, whose energy supply comes largely from material washed in from the watershed. The other is based on phytoplankton grown in the lake from recycled mineral nutrients; their energy supply is the photosynthesis occurring in the lake itself.

The two systems exist in all lakes, but their relative importance differs from lake to lake. Small ponds surrounded by woodlands fill quickly with outside material, and enormous lake basins may accumulate material for millions of years without filling up (e.g., Lake Tanganyika).

It seems reasonable to extend the basic picture developed for lakes to the oceans and seas. For relatively shallow, more or less enclosed portions of the sea, such as Long Island Sound, Chesapeake Bay, and San Francisco Bay, the extension works reasonably well.

There is a temptation to think of oceans as lakes grown amazingly large. Oceans have phytoplankton and zooplankton and fishes. There are locations where upwelled water produces immense blooms of plankton and yields of fish. But this biological recycling is a relatively local process. On a large scale the system is primarily controlled by external forces.

The *average* depth of the oceans is four thousand meters, or around two miles. There are changes in seawater density with depth over almost the entire ocean, so except in relatively small areas very close to shore, the entire ocean is covered with a seal of relatively low-density water, like a lake during a period of stagnation. No wind is strong enough to completely stir the depths of the ocean.

Near the surface, in both lakes and the ocean, particles of dead and decomposing material—body parts, feces, molts, skins, and scales—sink. As decomposition proceeds, these particles become fluffy masses of bacteria and shreds of tissue, approaching closely the density of the water around them. Fluffy sediment particles can sink for miles into the ocean depths, but the rate of sinking becomes ever slower as the sinking particles disintegrate, so they are essentially at the density of the surrounding seawater. Most of the nutrient minerals in these sinking particles diffuse into the surrounding water.

If a sinking particle is purely organic and can be decomposed by bacteria, it may disappear at the middle depths and never sink to the bottom at all. If, however, the particle contains a mineral core to act as ballast—a grain of volcanic dust, a bit of silica from a sponge, a carbonate platelet from a protozoan—it will not simply dissolve but will eventually reach the bottom. These ballasted particles also contain nutrients, either adhering to their surface or incorporated in their structure.[40]

How do these nutrients return to the surface?

There are two mechanisms for returning nutrients from the oceans' dark deep water to surface water. One is that there are some places in the tangle of currents in the sea where deep water is brought to the surface. The time between the sinking of a detritus particle and the return of the nutrient to the surface along this route is measured in centuries and millennia.

There is another, slower, but ultimately much more important cycle operating.

The watershed for a pond may be a single valley, but the watershed for the ocean is all the land on earth. All of the rocks of all the continents are washed in water heading for the ocean. This drainage is where nutrients come from, balancing the loss to the deep sea from the sinking of detritus.

At first glance this seems unreasonable. Why hasn't the land been exhausted of its nutrients in the last few billion years? It would have been, except for the fact that many of the rocks and mountains on land were originally formed under the sea. They are made up of the nutrient-rich sedimentary detritus that succeeded in sinking to the bottom and has accumulated over millions of years. This sounds like wild speculation, but it is much more certain than most supposed scientific facts, because we can find the fossils of deep ocean animals in the rocks of high mountains.

Oceans perhaps could be considered as very large lakes, but the basic picture developed for lakes has deflected research interest away from some more realistic insights. Communities on dry land involve more complex ecological interactions than those in the open water of oceans or lakes.

So far our descriptions have ranged from the entire earth to oceans and landscapes and lakes. We live on a planet with an abundance of water, a comfortable range of temperatures, and a reasonable distribution of surface minerals. It is because of this good fortune that life emerged. There was no inevitability about either the emergence of life or our own evolution. We have seen neither malice nor compassion in nature.

Ecological systems all transform solar energy into the energy of chemical bonds, and all organisms use the energy in these chemical bonds to maintain and reproduce themselves. Some energy-rich carbon-containing molecules, such as coal and oil, may be buried for millions of years. Others, such as the sugars in plants, are consumed almost immediately. Ultimately all the energy derived from the sun will be dissipated as heat. All the carbon atoms incorporated in energy-rich molecules by photosynthesis will be discharged as carbon dioxide, buried in sediments, or reused in photosynthesis.

Solar energy, wind and water, and astronomical cycles are sending streams of elements and energy flowing through all the organisms and the sediments, atmosphere, streams, lakes, and oceans. These big processes have an almost Wagnerian feel.

Can these gigantic systems be in any possible danger, even from the power of modern humanity? Generally the answer is no, with some important exceptions.

Let humanity do its worst! Rain will still fall, rivers will still flow, and there will still be storms and floods and droughts. There will still be photosynthetic organisms pumping out oxygen, nonphotosynthetic organisms using it and burning the products of photosynthesis. However, there is no certainty that any particular species or landscape will survive. The problems for humans are in the details. Will our particular species survive?

Some global properties involve relatively small quantities of matter—thousands of tons rather than millions. People can affect these, and they can change rapidly over periods of tens, hundreds, and thousands of years rather than thousands of millennia. The quality of air and of river water and the chances for survival of some landscapes and of many species, including our own, are in our hands.

2

How Do Species Survive?

POPULATIONS

With the exception of some changes in the atmosphere and in water quality, most practical problems of ecology are on a relatively small scale. Particular ecosystems, watersheds, and very often particular kinds of organisms are our concern.

Some broad generalities emerge from the facts of natural history. Some of these "laws" and "rules" are named after their proposers, for example, Bergmann's rule, Allen's rule, and Dollo's law. Bergmann's rule states that closely related animals tend to be larger in higher latitudes and sometimes at higher altitudes. This is perfectly plausible for birds and mammals given the dangers of heat loss in cold situations and the necessity for heat loss in excessively warm ones.

The closely related Allen's rule asserts that tails, ears, and other protruding structures of mammals are expected to be larger in warm habitats than in cold ones. Desert foxes have large ears, Arctic foxes small ones.[1]

These generalizations provide insight but are not useful in making specific decisions about environmental problems.

Dollo's law asserts that if a species becomes extinct, it cannot be reevolved. The process of evolution is in this sense irreversible. Although Dollo's law provides no guide to specific action, the irreversibility of species extinction is a powerful motive for conserving living species.[2] For managing nature, a more detailed kind of understanding is required.

Almost a century ago, a pioneer ecologist wrote that if we understand animals, "when we see a badger we might say: 'There goes a badger' in the same way we might say 'There goes the vicar.'"[3] The ecological niche of a species is analogous to its role or profession.

Ecological niches differ between species. To describe the ecological niche for any particular species we must know its relation to its physical environment and to other organisms—what it needs from them, what it might supply to them, and how they are affected by each other's presence or absence.

Any group of organisms of the same species that is going through the processes of birth and death and that can be considered as being in some way related can be called a population. Exactly what is meant by that is reasonably flexible. We might speak of all the Indian elephants on earth or tapirs on a specific island. The oysters or lobsters of Long Island might each be considered a population, as can the human residents of the United States, the redwood trees of California, the mosquitoes in a pond, the descendants of Thomas Jefferson who are now alive, and so on. All the individuals having some peculiar property (for example, all the white elephants in Africa) or suffering from a specific problem (all the cats in Manhattan that are allergic to milk) constitute a population that may be interesting.

Populations and communities of organisms are often defined as matters of convenience. Usually a biological population consists of a group of plants or animals and their descendants found in a particular place. This is a loose definition. Nevertheless, some statements about biological populations are absolutely and literally true.

It is literally true that all organisms that are now alive are the end of an ancestral chain extending to the beginning of biological time, four billion to six billion years ago. Also, it is literally true that the vast preponderance of lineages have not succeeded in persisting to the present. I'm not exactly sure what a "vast preponderance" is, but it certainly means that at least a million lineages that existed sometime in the last billion years have become extinct for every one we find today.

There are statements that I believe to be true but hold with less conviction. For example, past history leads me to believe that around 99 percent of the lineages that are now alive will be extinct by 10,000,000 A.D.

Persistence of a population depends on birth and death being approximately equal most of the time. A population must be able to increase in good times to make up for its reduction in bad times. "Good" and "bad" differ from species to species and situation to situation. If we understand the ecological niche of a population, we can make observations that will permit us to tell good from bad.

Even if we don't have detailed information about an ecological niche, we can generally tell how well a population is doing by examining its pattern of births and deaths.

There is a general question about births and deaths. Newborn plants and animals are generally smaller, weaker, and less capable of facing environmental problems than their parents. Also, giving birth and caring for young generally increases parents' chance of death. Why should so much effort go into producing new young organisms rather than preserving the old ones?

The answer is that no matter how superior the adults are compared to their own newborn young, old organisms do die. If they are not replaced, the population will disappear. For a lineage to survive, it must balance deaths with births.

While all organisms can die of accidents, why should many organisms become more subject to accidents and other sources of death as they grow older? They not only grow older, they age. Why do organisms age, and what determines how long they will live?

The age at death for most organisms depends both on properties of the organisms themselves and on environmental circumstances. In humans, social circumstances change the pattern. In the year 2001 in Afghanistan, one baby in four was expected to die before the age of five. In times of peace and even prosperity, within a single human family some die as babies, some in their teens, some live into their nineties. All die before their 130th year.

The oldest elephants live about as long as people, and some tortoises live twice as long. Dogs live for ten or even twenty years, but one master will outlive many dogs. Caged parrots often outlive their owners.

In most old humans some parts of the body retain all the properties of youth while others deteriorate with age. Some body cells do not show obvious aging. The skin of very old humans is not like the skin of a baby, but the lining of the mouth, throat, and intestinal tract does continue to slough off a layer of surface cells and produce an abundance of replacement cells until death. Bone marrow continues to produce red blood cells, and hair and fingernails continue to grow.

Other parts age more severely. After early childhood, human bones lose much of their capacity to recover from breakage. Concern with and capacity for sexual intercourse declines with advancing age. Sight and hearing are certainly not improved by age.

Some body parts are committed in irreversible ways as development proceeds. Vertebrate brain cells have interconnections that are vital to their function. If a brain cell is lost, as they are at a rate of several per second throughout our lives, no new cell can be inserted in that precise location with that precise set of connections. This does not necessarily mean that the brain's effectiveness is significantly declining, but it does mean that there are changes in the brain.

Some songbirds lose neurons and stop singing at the end of each breeding season but regenerate new neurons and learn new songs for the next season.[4] This has raised hope that other vertebrate brains may also regenerate.

Aging can mean a simple wearing down of the body's hard parts. This is apparent in teeth. We cannot regenerate teeth, nor do our teeth continue to grow, although our "milk teeth" are replaced.

Mammals may die because their teeth wear out or break. Old cows can live much longer if care is taken to avoid grit and stones in their feed or if they are given false teeth. But some animals' teeth are replaceable. Elephants carry hidden molars that grow forward to replace the ones that wear out. Sharks regularly replace all their teeth.

Sometimes the wearing away of a structure as it ages is vital to its proper function. For example, rodents' incisor teeth consist of two layers of material, an outer hard layer of enamel—often yellow or orange—backed up by a slightly softer layer of light-colored dentine. The incisor teeth continue to grow throughout life. In use, they work like self-sharpening chisels. The dentine wears slightly more quickly than the enamel, keeping the tooth edge sharp.

A rodent must eat food that provides sufficient wear to compensate for the growth of the teeth or the teeth will grow through the animal's palate and into its brain. So long as the balance between growth and wear is maintained, incisors remain at approximately constant length.

The size of organisms makes a difference in how susceptible

they are to various causes of death. Getting bigger as they get older lets them outgrow many sources of death. Baby animals are more easily taken by predators. Seedling plants are easily bitten off at the ground or dry out.

But size has its own problems.

Organisms transport materials through their surfaces and carry on most biological functions throughout their volume. A very large animal finds it proportionally more difficult to eliminate waste products and absorb food than a smaller animal of the same shape, because the surface area per unit volume is low.

A small animal finds it more difficult to keep warm than a larger animal of the same shape because the surface area, from which heat is lost, is large compared to the body volume in which heat is produced. Bats and mice huddle together to sleep; bears sleep alone.

It seems necessary for animals to keep an approximately constant surface-to-volume ratio in order to maintain function. This explains, in part, why the lung of a frog, a relatively small animal, is an empty sac while a cat's lung is a dense sponge.* Earthworms' intestines are straight tubes with a smooth lining, and our intestines are coiled and have a velvety pile on the inside. The pile in the lining of our gut increases the absorptive surface of our gut the way cotton pile increase the absorptive surface of a bath towel. The total surface area packed into our lungs or gut approaches that of a tennis court.

Larger terrestrial organisms have a greater mass of inert supporting structure in proportion to their musculature than do smaller ones of approximately the same shape. This explains, in part, the ponderous dances of circus elephants compared with

---

* Notice that many mice and moles are smaller than bullfrogs, but the bullfrog lung is a sack and the mouse or mole lung is a sponge. This relates to evolutionary history.

the vivacious movements of circus ponies. An adult cat carries its tail in a graceful curve, while a kitten's tail stands straight up. Holding a short tail up straight is much easier.

When wind pushes on a tree the tree acts as a lever, with the fulcrum at the ground. If the tree is tall enough, a relatively moderate wind can blow it over, other things being equal. As is so often the case in biology, other things are not equal. Tall trees generally have disproportionally wide trunks, using this greater resistance to compensate for the leverage associated with greater height.

Flat-leaved trees tend to drop their leaves before the full force of winter winds blows on them. Needlelike leaves are less resistant to the force of wind and usually are retained in the winter. Rhododendron bushes have flat leaves that curl into narrow tubes when it is cold. This reduces the danger from winter winds.

If immortality were merely a failure to die, there would be the horrible possibility of showing more and more of the symptoms of age without the possibility of relief in death, as in the Greek myth of Tithonus, the prince of Troy who was chosen as an immortal lover by Eos, the goddess of the dawn. This happens only in fiction or poetry.[5]

Some organisms can live a very long time. Bristlecone pines in the American Southwest are not very imposing trees, but some have survived for several thousand years. I have seen them clinging to cliff faces where the next rock slide may tear them out of the ground, or growing next to narrow paths that may be dug up and widened. Short of such violent events, each of them can probably live several more centuries at least. There are ground-dwelling molds and individual marine corals that have also survived for thousands of years.

Usually the young and the old are most susceptible to dying. The causes of death are often age-specific. In the United States, babies die of infections, nutritional problems, and accidents. Old people die of cancer, heart disease, and organ failure. The least likely to die are postadolescents, and when they die it is often due to accidents or suicide.

Why are death and age related? It is not obvious that age must carry with it the disabilities humans associate with aging. Aging occurs in organisms in which individual history leaves permanent scars. Teeth erode, skeletons carry marks of past breaks, and childbearing may permanently change a mother's figure.* If all the parts of an organism regenerate, time may pass but aging may be absent.

Some organisms do not age at all. In organisms that show no sign of aging, death occurs only from accidents or general deterioration of environmental conditions. A sign of possible immortality is a death rate that is constant with age. In the freshwater polyp, *Hydra,* there is no skeleton. Temporary starvation results in a smaller animal with lower demands for food. There is no brain, so the subtleties of neuron-to-neuron connections do not matter. The animals are capable of regenerating all lost parts and were therefore considered to be potentially immortal.[6] An assemblage of these animals was maintained in my laboratory for several years to directly test their potential immortality. No changes of any kind occurred in them during this period.[7] While individual cells in the animals vanished and were replaced, the hydra themselves were unchanged and were, in principle, immortal.

Groves of aspen trees often consist of genetically identical trees connected through their roots. Such a grove can persist for

* This is in part environmental. Several decades ago physicians tried to keep down weight gain in pregnant women, and the effect of pregnancy on their figures was less than it is today. The change is rationalized in terms of infant health, but I do not have actual statistics.

millennia. They originated from one tree whose roots have grown out and sprouted new trees. They can be thought of as one gigantic, potentially immortal individual.

Among organisms of tremendous age essentially all, but not quite all, of the offspring die young. The puffball on the forest floor is the fruiting body of an underground meshwork of fibers that may grow very large and may be hundreds of years old. I don't know how many visible puffballs it produces at any one time, but a single puffball ejects literally billions of spores. The number of puffballs does not increase from year to year. On average, each meshwork of fibers and its fruiting bodies produces only one descendant. The chance of one spore surviving to produce a new fiber meshwork with new puffballs is much lower than my chance of being hit by lightning on the day I won the top prize in the lottery and was called to grand jury duty. The very few that do survive to reproduce replace all those that die.

There are probably many extremely ancient individual organisms living unnoticed all over the world. In Michigan there is a single gigantic mold that exists as thin fibers growing through the soil over several square miles. It is one organism but has essentially no solid body, except when it produces mushroom like reproductive bodies. Its age is not known but is probably measured in thousands of years. Large corals also may live for thousands of years, so that their great-grandparents would have coexisted with a glaciated North America.

The symptoms of age vary. A tree may have fewer leaves and a relatively larger trunk as it ages. All insect wings are composed of already dead tissue (like our fingernails and hair), so that old butterflies have frayed and broken wings. Old dogs develop graying muzzles. Most old men are "wonderfully weak in the hams" (as Hamlet said to Polonius), distressingly forgetful, sexually incompetent, and intellectually disinterested.

Different parts may or may not be replaceable. Birds' wings carry feathers, which cannot be repaired once they have formed, but they can be replaced. The wings of bats are layers of skin, like the webbing between the bases of our fingers, but much larger in proportion. These do not wear out.

The animals we see in nature are almost all in their prime. At the first sign of weakness aging animals in nature die—perhaps eaten, swept out to sea, or crushed by a falling stone or a hiking boot.

In some organisms the time of death is built into their anatomy and development. Many species of mayflies, belonging to an order justly named Ephemeroptera, have no capacity to feed after they reach sexual maturity. In several salmon species the males that fight their way upstream to breed lose their intestines as they enter fresh water. These doomed males are reduced to a transport system for sperm and no longer have any future of their own.

In short, when organisms die and what actually kills them are surprisingly varied.

If total births during a time interval exceed total deaths, a population increases. That is obvious, but to say that there were twenty-five deaths and seventeen births this year has very limited meaning. Obviously the population has declined by eight individuals, but this has an enormously different meaning if the population size was twenty thousand as contrasted with fifty. We may learn more if we consider birth and death as rates. Then the same figures would indicate low birth and death rates in the larger population but dangerously high ones in the smaller.

Patterns of death and birth vary among organisms.[8] Some animals, such as oysters and codfish, may produce literally millions of young at once, but almost all of these young animals die within hours. The adults usually survive to breed again.

Many birds, after they have left their nests, die at an almost constant rate, independent of age. In most species of small birds this rate is heavy enough so that the average bird lives only a few months.[9]

Some large seabirds, such as the albatross, may live for decades but produce only one or two young every other year. Termite queens produce a few young a minute and continue reproduction for ten years or more. Salmon produce thousands of fertilized eggs as their last act before dying. Armadillos generally have multiple births, often quadruplets.[10]

People usually have one child at a time but can produce as many as seven. The life expectancy of newborn people varies with social conditions but in prosperous countries may reach more than seventy years.[11]

Some organisms do not reproduce at all until just before their death, when they produce a large number of young. An extreme case is found in those cicadas that spend either thirteen or seventeen years underground with their tubular stylets inserted in a growing plant, drinking sap. After the correct number of years has passed they emerge and produce loud chirping songs for a few hours or days of mating and egg laying and then die.[12] Some bamboo plants also crowd their reproductive life into their terminal days after many years of vegetative growth.

In codfish, millions of fertilized eggs are produced each breeding season by every mature female. In most years essentially all of them die. To survive the first day and a half of life, newly fertilized codfish eggs require that the turbulence of the sea be enough to keep them from sinking too deeply and not so much as to disperse them into water of the wrong temperature or salinity. Also, the population of tiny predators, including copepods, arrowworms, small jellyfish, and other fish larvae, must be small enough so that not all of the cod eggs are immediately eaten.

In a few days baby cod develop a gut and a mouth, but by that time they have almost exhausted their stored yolk reserves. Once they have a mouth they are tiny predators, but with only enough strength to make five, ten at the most, lunges at food objects. If these succeed, the probability of staying alive for even a week goes up enormously. If the baby cod do not succeed at these first predatory efforts, they will starve. Note that if their mother had endowed each of them with more yolk, she would not have been able to produce as many of them.

In most years essentially no baby cod survive. In some few years, at irregular intervals determined by chance coincidence of favorable factors, the survival rate up to a size of three or four centimeters may be as high as .05 percent. By that size the baby cod have outgrown many mortality sources and can swim strongly enough to follow food particles, escape small predators, and even stay in the right kind of water despite weak currents and turbulence. The bigger they get the greater their chances of surviving another week, month, or year. A one-year-old cod has a very good chance of living another five years or even more. Those few years in which survival of baby cod is a trifle higher produce "good year classes." In the days prior to World War II, before the development of massive floating fish-catching factories, particular year classes dominated the Icelandic cod fisheries for many years.

Long Island oysters have much the same pattern as cod. There is an enormous number of young, each with very short life expectancy, but life expectancy grows longer the older they become.

In most mammals and birds there is a high death rate among newborns, but nothing as severe as that in cod, oysters, puffballs, and all the organisms that produce a tremendous number of tiny young at one time. Life insurance premiums for cod and oysters should, in fairness, decline with age.

The patterns of a large number of tiny young or a small number of well-provisioned or cared-for young are as characteristic of particular species as their anatomy. Some organisms produce a relatively small number of large young, which are either cared for by their parent, as in mammals, or endowed with generous energy reserves, as in coconuts. Even among mammals there are differences. Opossums may have ten or twelve young at once, elephants only one.

When environmental conditions are poor the longevity and number of young may decline, and when conditions are excellent the number of young born and the rate of survival go up. Nevertheless, the basic patterns persist. They are the outcome of the evolutionary process of natural selection. Why are they so different?

We can answer that in terms of details of each kind of organism's ecological relationships. We can also develop a kind of general understanding, if not an actual answer, by considering that what the evolutionary process is selecting for is ultimate success at the transmission of genetic material.

There is a temptation to assert that evolutionary success is measured by actual reproduction. We know this cannot be true since there are evolutionary lines in which low reproductive rates have a selective advantage over organisms with higher individual reproductive rates. I have not made a complete survey of how many times this has happened, but it is clear that it occurred in the evolutionary line that went from primitive primates to humans.

Most individual organisms fail at reproduction. We all know of trees that have never borne fruit, of puppies that died or were neutered before becoming parents, of small fishes preserved in sardine cans, of seeds that were eaten before they could grow up and produce more seeds. Each kernel on an ear of corn is a potential corn plant. Eating an ear of corn ends its cycle of birth and reproduction. In fact, the overwhelming preponderance of organisms all through the history of life have not been able to continue the passage of life to offspring. But it is quite obvious that every individual organism now living is the result of a continuity of generations of parent and offspring that has been going on for literally billions of years, back to the origin of life itself.

Many organisms succeed at reproducing but never encounter their young. The spores of mold can travel hundreds of miles on the wind before touching earth and germinating. More surprising are the organisms whose young move in time. Many kinds of small aquatic and marine crustacea have two modes of reproduction. They can produce eggs that hatch immediately, thereby entering the same population as their mothers. These eggs are produced when food and physical conditions are very good. They permit rapid and immediate increase. Often they do not even require fertilization. But when environmental conditions for individual *Daphnia* and copepods deteriorate, they may produce eggs that require fertilization.[13] The mother may produce a chitinous envelope for these fertilized eggs. Even casual examination of the mud at the bottom of a lake or pond will demonstrate that the chitinous cases of cladocerans' eggs are conspicuous and common. Copepods also deposit cased eggs, but the cases are less conspicuous.

Using isotope dating and extracting the eggs from thin slices of sediment, it has recently been found that copepod eggs can persist in sediments for more than three hundred years and still hatch.[14]

Living cells can die. Many of mine will die today, but, with luck, I will be alive tomorrow. The death of cells does not imply the death of organisms. Individual organisms can survive the death of their component cells or even the loss of some of their organs. Populations can survive although many of their members have died, and species can survive even if some of their populations die. Individual members of nations, tribes, and cultures may die, but the social units may live on. In fact, many peoples have rallied around ideas such as "It's good to die for our fatherland."*

Sometimes the death of an individual and the death of a tribe can coincide, but that is most unusual. When Ishi, the last of the Yahi tribe, died, one population of humans died with him, but humanity as a species had not died.[15]

At the Cincinnati Zoo on September 1, 1914, a pigeon named Martha died. This was not only sad. It was tragic because that pigeon was the last passenger pigeon in the world and its death was the death of the species itself. What had been the most numerous bird in North America a hundred years earlier was now forever gone.

It is of great importance to realize that there is no evidence whatsoever that species or populations age the way you and I are aging. Individuals in populations suffer from aging, but the population itself does not. Species that become extinct do not do so because of great age.

INDIVIDUALS AND POPULATIONS

Peter Medawar, a Nobel-laureate developmental biologist, wrote a highly mathematical book and gave a copy to his colleague Conrad Waddington. When they met a few weeks later, Medawar asked Waddington how he had enjoyed the book. Waddington said he had loved it. Medawar then asked how that could be since Waddington had almost no knowledge of math-

---

* A comment attributed to General Patton is: "Better than dying for your country is to make the other son of a bitch die for his country."

ematics beyond elementary algebra. Waddington replied: "I hummed the hard parts."*

Parts of this section can be hummed. It introduces two useful concepts from the theory of ecology: the rate of population increase, R, and the age-specific reproductive value, V_x. There is a very good computer program and textbook that does all the arithmetic for calculating these values along with such things as mean life expectancy and even the likelihood of extinction during a time interval given statistical variation in these various values.[16] You may therefore develop an understanding of these things without worrying about actually making calculations.

R is evaluated from the probabilities of birth and death. Any cohort of organisms can in principle be observed from the time that they are of age zero until the last of them has died. We can make a graph of survivors against time, showing what percentage of the organisms is still alive at what age. This is a survivorship curve. The development of survivorship curves deserves some attention.

It is enormously inconvenient to collect a cohort of newborns and follow them through their lifetime till the last one dies. Not only are there problems in their maintenance, but also many organisms live longer than any research grant, and some, like elephants, trees, lichens, and tortoises, live longer than most investigators.

What we want is a survivorship rate, not an absolute number of survivors. We estimate the survivorship curve by examining the rate at which organisms of known age are currently dying. The techniques vary with different organisms. For many organisms, a survival curve can be estimated from data that can be collected today.

We cannot determine the age of all organisms, but we can

---

* I heard this in a lecture by Medawar around twenty years ago.

tell the age of some of them.[17] In an early classical study an investigator collected the skulls of mountain sheep. He could tell the age of death of each skull by counting rings in the horns. The relative abundance of dead at different ages could be used to make a survivorship curve.[18]

It is possible to tell the age of the trees in a forest by taking narrow cores of wood from the trunk and counting the layers of wood. The ear bones of many fishes show annual growth rings, so a sample of captured fish gives an estimate of the abundance of different ages in the population, from which an estimate of age-specific death can be made.

A survivorship curve can be made either by measuring today's age-specific death rates or by starting with a cohort of newborn individuals and following them through their life history. The results may differ. This may be most easily seen by considering an example of human survival.

My mother was born in 1903. She survived what were then major killers—diphtheria, whooping cough, the flu epidemic of 1919, and a series of wars. She lived to the age of eighty-six and died of a torn aorta. A baby born in 2003 would be immunized against many of these sources of death and would have a much higher probability of long-term survival than children born in 1903. However, it is quite possible that the babies of 2003, not being winnowed by childhood killers, might be generally less tough and succumb to more causes of death in late middle age. Also, new viral diseases, sexual diseases, and poisons are available to mold the survivorship curve of a twenty-first-century cohort.

A survivorship curve derived from the age at death of humans in America during the year 2003 will have a different shape from either a cohort followed from 1903 or a cohort followed from 2003. Also, survivorship from African data, European data, and American data will differ. Nevertheless, the approximate shape will be the same—dangers near birth, an increase in death rate at advanced age, and high survival during the prime of life unless war or some new disease intervenes.

In a similar way, by observing today's organisms we can note how many newborn organisms they produce and at what age. This produces a fecundity curve, births per living organism against age.

These curves of survivorship and fecundity with age for any group of organisms can be used to evaluate both R and a number called *reproductive value,* a theoretically rich descriptor that can be used to develop analytically valid laws of some aspects of ecology and also of evolution.*

Age-specific mortality and reproductive rates can be measured for the specific organisms of interest. These can be entered into the equation in appropriate locations, leaving R as the only unknown. It is impossible to solve these equations analytically for the remaining unknown term, R, but once we decide on the reproductive value at birth (1 if for females only, 2 if for both males and females) we can solve for R by substituting test values into the right-hand side of the equation until we find the value of R that makes the equation balance. The value R is the rate at which the population supplying the life table data would increase ($R > 1$) or decrease ($R < 1$) if the survivorship and fecundity values stay the same. The increase or decrease is per year or day or hour or whatever time unit is specified by x, the age increments used in the calculations.

If conditions are right, all populations can increase, almost like money in an interest-bearing savings account. If you have an account with a 6 percent annual interest rate, we could represent the amount in the account one year from now as today's balance multiplied by 1.06. Two years from now you would have

---

* See the Appendix for a discussion of how to determine the value of R.

today's balance times 1.06 times 1.06, which could also be written as today's balance times $(1.06)^2$. The R value for your money is 1.06. After sufficient time, and with no withdrawals, this would result in your bank account becoming enormous.

The increase of populations differs from the increase of money in the bank. In a bank account all the dollars are equal. In a population, some individuals contribute nothing to population increase, others contribute a great deal. Old ones no longer breed, and some of the young ones wait a while before they have young of their own. Also, in an honest bank account dollars do not leave without your permission. In a population individuals can leave by dying or even by wandering off. And the interest rate is not equal for all organisms.

Despite these difficulties, if conditions for the organisms in the population persist long enough, it can be proved that a single number can represent the population's increase rate, R.[19] When $R = 1$ the population is just maintaining itself. Even a small deviation from this value implies rapid changes in population size. $R = 1.03$ per year doubles the population in thirty years.

All populations of organisms can increase exponentially if circumstances are just right, but circumstances are never just right for very long. Very often $R < 1$ and the population declines.

For all living species, $R = 1$ on the average. Organisms replace themselves generation after generation. They and their ancestors have done so since life began. If they did not, we would not be concerned with them further, except if we were paleontologists who must puzzle out how they lived and why they disappeared.

There are several significant features hidden in the concept of "average." The evolutionary line that includes mammals and mammal-like reptiles began with one population sometime around four hundred million years ago. They have diversified.

They started as crawling animals vaguely resembling a crocodile. Their descendants include modern crocodiles but also humans, bats, and kangaroos. Certainly this is an increase not only in diversity but also in abundance.

No increase in abundance from generation to generation would have been noticeable. Instead of R = 1 we might have had R = (1 followed by one hundred zeros and then another 1) (R = 1.00 . . . 1).

The word *average* covers up the fact that during any particular short time interval for any population, R is very likely not to be equal to 1. Many organisms have a discrete reproductive season. Flowers and their seeds appear in the spring, lambs are born in the spring, and small birds nest and breed in the spring. For all of these the spring represents a season of increase and the rest of the year a season of slow decline. Nevertheless, on a reasonably long-term average, R = 1 for all of them.

Reproductive value, V_x, is defined as the equitable current value of the loan of a life to an organism age x born at time 0.[20] This is not immediately intuitively clear, but it is important. Three terms enter into the evaluation of an organism's reproductive value: (1) a population rate of increase raised to the power x, which takes account of inflation or deflation since the time of birth of the organism age x; (2) the inverse of the survival rate to age x, to take account of the fact that the lower the survival rate, the higher the reproductive value at age x must be; and (3) a summation of how many young an individual of age x is likely to produce in the future, with each age's production corrected for inflation (this is a sum of the terms in a vector consisting of the fecundity of the living animals age x and older, corrected for their probability of surviving to each of those ages). These can be combined to produce a data-rich equation from which V_x, the age-specific reproductive value, can be calculated.

The guiding image for thinking of reproductive value is that of bank loans. The members of a group of organisms born during some short time interval have each been lent a life. Unless those lives are returned with interest, the lending process cannot continue. If the population remains constant in size and if none of the organisms dies before reproduction, then it is sufficient for the continuity of the process that each organism in the cohort produce precisely one newborn organism before its death. This is its replacement.

It is essential in the business of lending money that the greater the likelihood of default on loans, the greater the interest rate the bank must charge. The greater the number of borrowers that have defaulted, the greater the return that must be made by those who have not defaulted. If some organisms die without reproduction (i.e., abscond without paying their debt), the number of newborns produced by the survivors must be correspondingly greater or the process of "loans" will cease.

Continuing with the image of a bank loan, the value of a loan decreases as it is paid off. Also, if there is inflation of currency, the temporal value of the loan must take that into account so that the total return to the bank from the cohort of loans has the same value in the context of the total money supply as it did when the loans were made.

This discussion seems forbidding, but taken in steps it is extremely simple. It is so powerful theoretically that it is very much worth the trouble of rewording.

Assume a cosmic lender of lives that does not make a profit but must maintain an equitable return if life is to continue. If the value of a life at birth can be taken as unity, this value must increase with age in response to defaulting. The reproductive value of the survivors must be correspondingly increased in response to the prereproductive mortality rate.

Repayment of the loan begins once the age of reproduction

is reached, thereby lowering the loan's residual value. Think of how the current value of a mortgage loan declines as the loan payments are made.

A correction term must be introduced into reproductive value because the return of one life for one life given is not equitable if there has been inflation or deflation of the population size during the lifetime of the organism. The inflation rate of a population is given by R. Notice that a mortgage rate takes the rate of inflation of money into account.

To see how reproductive value is used, imagine that you have contracted to populate a particular island with colonists and you will receive a cash payment proportional to the number of colonists living on the island fifty years from now. For the sake of political correctness let us assume the colonists are cats, not people. Also, let us assume for mathematical simplicity that we need not add any males to the island—either the cats in question reproduce without males or the services of males come free. We are speaking only of female cats.

How much are you willing to pay for cats of different ages that can be transported to the island? The best way to calculate the relative amount that you would be willing to pay for a cat of a given age would be to calculate the cats' reproductive value.*

Kittens will not be worth much, since many accidents may intervene between the time you purchase the kitten and the time it begins to contribute to the population. Old cats will have suffered the aging process and will reproduce slowly if at all. Top price will go to adolescent animals since they will immediately contribute to the increase of the population. Also, you would pay more for them if they could be delivered immediately, so that you get the advantage of their natural increase.

---

* Obviously that depends on the condition of the cats, but let's assume that all the cats offered to you for purchase are in fine condition for their age.

This is equivalent to valuing a present payment above a future one in money lending.

To get a stronger sense of what is involved here, we can ask: If conditions do not change, how much will a woman who produces twins at age nineteen contribute to the human population a generation from now compared with producing a larger number of babies starting at a higher age? In fact, having twins at nineteen is approximately equal, in terms of contribution to later population size, to having five children starting at age thirty.

Birth and death rates, unlike tooth structure or blood type, are very sensitive to immediate environmental conditions. Hungry animals or thirsty plants do not survive well, nor reproduce abundantly.

If conditions for a population are particularly beneficial, survival and fecundity will both be enhanced and R can exceed 1, at least temporarily. This is what permits small groups to rapidly become abundant. Fewer than twenty rabbits, of the right kind, and their descendants essentially took over half of Australia in less than a century.[21]

This is also what permits such animals as elephant seals, Pacific otters, and American bison to recover rapidly from tiny populations on the verge of extinction. Bison were almost extinct a century ago, but bison meat is now a standard hamburger ingredient in western North America.

These tiny populations that became enormously abundant during their period of increase were growing at close to their maximal possible rate, referred to as R_{max}. For every kind of organism there is a set of circumstances at which this maximal rate can be approached. Precisely what these circumstances are varies from organism to organism, except we know that for all organisms crowding by organisms of their own kind causes deterioration in conditions for each of them. R_{max} occurs only when the population size is very small or the animals are not at all crowded.

More than a half century ago a graduate student demonstrated that smaller organisms have a higher level of R_{max} than do larger organisms over a tremendous range of classifications and body sizes. He suggested that the relation between R_{max} and body size is an evolutionary response to the fact that smaller organisms are more frequently confronted with complete destruction of their habitats or elimination of local populations. This requires them to have the ability to increase rapidly when favorable conditions become available. Larger organisms are less frequently confronted with the need to invade new habitat.[22]

Generation time—the interval between the birth of a mother and the birth of half her offspring—is also inversely related to body size. The relation holds from whales to viruses to bacteriophages. Phages are tiny particles living within single bacteria. The smaller the phage, the higher its R_{max}.[23]

There is no mathematical reason why these properties should be related to adult body size, nor is it obvious from biochemical considerations. The biochemistry of organisms is pretty much the same, and a large organism, in principle, can increase as rapidly as a small one.

There are general associations between the fecundity, survivorship, and body size of organisms. If the survivorship curve shows a very high rate of death early in life, it will usually be accompanied by very small body size at birth. If the survivorship curve shows greater survival for the young, the young will be larger compared to the mothers. The concepts of R, R_{max}, and V_x permit comparisons and inferences about ecological systems, which are sometimes of practical importance. They provide guidelines for the management, exploitation, conservation, and eradication of populations.[24]

Elimination or addition of an organism of high reproductive

value makes a greater difference in the subsequent history of a population than elimination or addition of an organism of low reproductive value. For fish that produce a large number of eggs and spawn them out to fend for themselves, the curve of reproductive value against age starts at 1 at $x = 0$ and then increases enormously because of the defaulting by so many of the young animals. Immediately prior to the first breeding V_x may equal several hundred. With the first breeding V_x drops precipitously.

A standard way of preventing fishing from destroying the population of fish is to limit the number of fish caught. It also is possible, while holding the yield constant, to lower the impact of fishing on the fish population by careful choice of the kind of fish taken. By adjusting the mesh sizes of nets it is sometimes possible to catch only large fish that have passed through one or more reproductive seasons, ensuring that the animals we remove are of relatively low reproductive value. We can also set the fishing season to follow the time of breeding.

Relative reproductive value is not the only consideration in designing a fishery, and some of these considerations result in poor choices from the standpoint of ecology and species preservation. In fish such as salmon, whose market value drops rapidly at breeding, other adjustments must be made. The caviar and sea urchin fisheries take mothers with eggs. In both cases the reduction of the exploited populations has been very severe.*

Sometimes particular biological properties permit fisheries to try to minimize damage in other ways. For the famous stone crab restaurants of Florida, fishermen break the large claws off

* The luxury market for sea urchin eggs is for Japanese restaurants. The populations of urchins in the seas near Japan have been destroyed. Japanese buyers now get sea urchins from the coast of Maine. The Maine urchins are becoming scarce.

living crabs and then return the crabs to the water, to presumably regenerate new claws. I do not know the evidence that the crabs survive this well-meant operation.

In 1947 Evelyn Hutchinson presented a lecture on the use of the famous snail-based purple dye by pre-Columbian Mayans. Apparently the only way to explain the variegated colors on the purple cloth was to assume that the thread from which the cloth was woven had been passed through snail's liver an inch at a time. He felt that the way this was done was to carefully crack the snail shell, pass the cotton through the liver and then return the snail to the sea to heal.

We may want to destroy or reduce a population that is declared a pest or a danger. In that case the attack should focus on the organisms of highest reproductive value. For example, it has long been known that the number of rats in a modern city approximates the number of humans, and if public sanitation deteriorates, the rats can outnumber the humans.

Rats are usually attacked by placing poison baits and traps in places rats pass. This procedure selectively destroys itinerant males. The animals of highest reproductive value—the nesting mothers—may not be damaged, nor is their care of the young disturbed by excess attention from males. The effect of this sort of policy is that large numbers of rats can be caught and displayed, demonstrating that the eradication company is doing its job, but the reduction per dollar in the rat population is minimal, ensuring that the eradication company and the rats both can continue.

Similar arguments can sometimes be made in the context of disease eradication—often the serious transmitters with high reproductive value are situated in such a way that their elimination is extremely difficult.

The idea of reproductive value is built on obvious properties shared by all organisms. Loss from a population of an organism with high reproductive value will make a greater difference in the future genetic properties of the population than loss of an organism with low reproductive value. Natural selection matters most when it affects high-reproductive-value organisms. This permits us to explain, in large part, why predators often give the illusion of acting like good conservationists when they capture their prey. In many cases the only prey a predator can capture are the weak, very old, or very young. These are the ones with low reproductive value and make the least difference to the prey population.[25]

SPECIES DIVERSITY

The number of different kinds of plants and animals that you can see varies from place to place. There are up to four hundred tree species per acre present in the rain forest, around forty in a New England hardwood forest, and perhaps ten in a forest in Maine.

Where light can reach the ground—for example, along a road or riverbank, or where a large tree has fallen—a tropical rain forest is often a living green wall of bushes and small trees matted together by vines. Once through the green wall, in forests that provide complete shade, it is too dark for plants to grow up from the forest floor. There are very few bushes, vines, or young trees. It is like walking into a cathedral. But forests differ. Some northern evergreen forests and some bamboo forests are too dense to walk through.

In the springtime in northern hardwood forests all or most of the trees are likely to blossom simultaneously, making a fine show. In the tropics blossoming time differs from species to species. When a tropical tree blossoms it is likely to be showier than most trees in the temperate zone.* When they have no flowers many tropical forest trees look alike and have big shiny leaves, reminiscent of plants that grow in the rich moist warmth of dentists' offices. Growing on some of the tropical trees are bromeliads (members of the pineapple family), orchids, and ferns, which make the lichens, algae, and mosses on the northern trees seem insignificant.

Not only forests show dramatic differences in the number of visible species. On the tidal rocks of New England there may be twenty visible species of fish and invertebrates and five or six species of algae. On the reef at Eilat in the Red Sea there are dozens of species of highly visible and differently colored corals, algae, and sponges. There are also crabs, worms, and more than a hundred species of fishes.[26]

These differences in species diversity require explanation. It is usually relatively simple to explain why a particular species is or is not present in a particular place, but attempts to study, explain, and interpret diversity in a general way are difficult and sometimes controversial.

Approaches to the study of diversity vary, differing in their foci and assumptions. Even the word *diversity* is used in many ways and will need clarification as we go along.

Global species diversity refers to all species on earth. Usually *species diversity* refers to the number of species of some general group, say birds or trees or fish, found in some relatively small geographic region such as Hawaii or New York or a specific coral reef or forest.

A coral reef, a tropical rain forest, and the great grassy plains in Africa are rich in large, fascinating, beautiful organisms. The Maine woods, the Sinai Desert, and most city parks look depauperate in organisms—particularly during daylight and if one

---

* Chestnuts, locusts, fruit trees, and magnolias can match most tropical trees.

insists on large animals and birds. What the diversity patterns are for soil arthropods, molds, or bacteria is not known.

It is not really clear why or how biodiversity is to be measured. Do you simply count species in a sample? This is useful for some theoretical purposes, but it has the weakness that a rare single flower carries the same weight as a herd of elephants.[27] Also, diversity should properly include the myriads of inconspicuous organisms in the soil, the parasites, the fungi, and ultimately the bacteria and viruses. No one is in a position to make such a count anywhere on earth.

If you do not include all the species in a place, how do you decide which ones to include? On one occasion we learned something of the natural history of one species of fish from counting just the number of species of fish in different local regions on a coral reef. High diversity was seen near places where cleaning wrasses, little fishes that removed parasites from other fish, had set up territories.[28]

Perhaps you can learn something more important if you weight the observed numbers in some way. There are at least thirty different ways of weighting observations for calculating diversity.[29] Why should different species be weighted in different ways?[30] There is no general answer.

In short, aside from the fact that the concept has tremendous popular appeal and lends itself to powerful and even poetic prose, why is diversity important?

Questions of biodiversity and invasive species have engaged massive popular concern. It is important to distinguish between invasive species and mere pests. For example, in 1975 I planted a patch of Jerusalem artichokes in our yard. Starting in 2000 occasional Jerusalem artichoke plants appeared in a vegetable garden around a hundred feet from the original patch, and in the spring of 2001 dense growths of Jerusalem artichokes burst out all through the vegetable garden. My wife referred to this as "invasive," which it is, but since the Jerusalem artichoke is indigenous to America, it cannot be called an invasive species in the legal sense used in President Clinton's edict against invasive species.[31]

The problem of species diversity has generated a great deal of argument, a large number of dedicated laboratories, and even considerable legislation, both in America and around the world. There are many volumes on the subject. There are entire institutes of biodiversity. An encyclopedia of biodiversity in four volumes has recently appeared.[32]

Basically, I remain somewhat unconvinced by much of the enthusiastic fervor that motivates studies of diversity. At this point I am in a minority. I will present my view as clearly as I can. It will be evident where dissenting opinions occur. Is it obvious that they lie on the edge of ecological thinking, or are they important because they engage popular concern, or do they have major practical consequences?

One way to investigate the role of diversity in ecological thinking is by examining how and where the term *species diversity* is used. A number of colleagues and I analyzed the frequency of use of the word *diversity* in research journals, designed for fellow scientists, and in popular publications, designed for the general public, up to 1996.* We found that *diversity* was much less frequently used in writing for fellow scientists than for the public.[33] Why the popular concern and why the apparent lack of actual research concern? I believe that the use in scientific writing has increased since 1996, but I do not know if this is due to scientific advance or to focusing of funding efforts.

But the questions presented by society to be answered by a particular science need not necessarily make for good science and need not even make sense. German government officials

---

* This was an exercise in a graduate student seminar in the Ecology and Evolution Department, State University of New York at Stony Brook.

prior and during World War II asked anthropologists and sociologists for scientific demonstration of their curious theories of racial superiority. Even brief examination of German anthropological journals from this period demonstrates that the anthropologists did their best to comply—to find plausible-sounding answers to nonsense questions because the people asking the questions controlled the means for doing "science." Opposition to these controls sometimes resulted in death and often resulted in poverty and exile.

The science of genetics in the Soviet Union was similarly corrupted to support erroneous ideas of biology. Specifically, there was an espousal of Lamarckism—the theory of general inheritance of acquired characteristics—because this seemed more in line with Marxist notions of history. To continue a career in standard genetics involved subterfuge and heroism.

Severe reprisals for disagreement with official policies indicate that the subject is being taken very seriously by those in authority. This may extend well beyond science. Poets in the Soviet Union were risking their lives before large audiences. American poets are at most risking their livelihoods before small audiences in academic award committees, university classes, and experimental theaters. These are extreme examples.

It is easier to gain support for research if there is agreement between the supporting agency and the scientists on what questions are valid and important. We can, for example, expect differences in support for study of the effects of global climate change if politically powerful figures accept the reality of global climate change, or at least accept the legitimacy of the scientists who are informing them about global climate change, than if they do not. Detailed presentations of the case for climate change were prepared by the congressional Office of Technology Assessment more than ten years ago.[34] Apparently many political leaders were not convinced; with a change in political control of Congress the agency itself was eliminated. From this I infer that any subject-matter-focused science that requires financial and other support from governmental or fiscal agencies is required to present as clear and solid an argument for its significant public assertions as it possibly can. It may not help, but it is appropriate to make the attempt.

Sometimes, for the sake of gaining support for research, we must make assertions that are not quite as definite as we would like, but when we do that we must state what we are doing and not try to paper over our scientific weaknesses. Sir Ronald Fisher, the great pioneer in application of statistical methods to biological problems, wrote that tobacco could not properly be blamed for causing lung diseases or cancer. He was employed at the time at Rothamsted Agriculture Station, which was partially supported by the tobacco industry.

Simplifications sometimes lead to theories about metaphors. These can be appealing, but may also be dangerous. For example, in one theoretical analysis of species diversity, it was asked how populations of different species are packed into a community. Species packing is a tempting metaphor, and there have been many theoretical formulations of species packing.

The greater the species richness of a region, the more species have presumably been "packed" into it. If more species can be packed into a high-diversity region, does each of the species have a narrower ecological niche than species in a low-diversity region? This sounds as if it is making some sort of sense until we notice that the idea of the ecological niche has been transformed from a description of a species' role to something with geometric properties, as if it were a niche in a wall.

Consider the idea of packing in a literal way. The word *packing* has at least two meanings that can cause confusion. Neither of these meanings is derived from biology. Packing may mean placing objects inside a wrapper or container without altering them, or it may mean squeezing them into a container of fixed size. Contrast the packing of glass tumblers with the packing of pillows. Tumblers may be packed into a container, but if they press hard against each other, they shatter. Pillows can be squeezed into a container and fluffed up again when they are taken out.

The possibility of either type of species packing presupposes some sort of container into which things are packed. Is the inferred container equivalent to an environment? Does an environment have fixed geometric dimensions, like a wooden box? A pond, a log, and a rotting fruit can all be thought of as containers. However, environments can be defined in ways that may have more meaning to investigators than to inhabitants. Are study areas such as the La Selva Research Station in Costa Rica, the country of Costa Rica, or all of Central America in some sense meaningful packing containers?

Species-packing theories are concerned with species in an implicit container. The boundaries of the container are usually considered to be permeable to the flow of resources and energy but, unless specified, impervious to the passage of organisms. These are important assumptions.

The formal theory of species packing originated from the equation systems of Lotka and Volterra.[35] Their equations and some of the systems derived from them are available in most elementary texts and in reviews, but fortunately, most of their conclusions can be summarized in simple language.[36]

The mathematician Volterra, together with his biologist son-in-law D'Ancona, considered several species competing with each other for resources in a space through which resources flowed at a constant rate.[37] They assumed that if the flow of resources and the physical conditions did not change, any single-species population would remain numerically constant after population equilibrium had been achieved.

They assumed that no resource is in infinite supply. If organisms of different species have similar resource requirements, there must ultimately exist an inverse relation between the number of species and the number of individuals per species that can coexist. It helped to think of species abundance as being directly translatable to rate of consumption of a temporally renewable resource.

Organisms can affect each other by reducing the availability of resources. They can also alter the chemical or physical properties of the local environment. These alterations are often referred to as *crowding*. Yeast cells secrete alcohol that affects other yeast and bacteria. Beetles and mice change the odor of the environment. Mussels and barnacles physically crowd each other. If organisms compete for resources, any individual organism will negatively affect the well-being of all the others in the designated space, as measured by birth and death rates.

Think of two species and a single space. Assume that either species alone could maintain a population in our space. Now introduce members of both species into the single space. This is packing two species together.

There are basically three possible outcomes. The first is that the two species may be so different in how they use the space that the populations of both can survive. For a crude example, what if the space contained a pond and a forest and we introduced a sparrow and a carp? We would not expect the bird and fish to crowd each other.

Another possibility is that individuals of both species are more sensitive to the presence of the other species than to the presence of conspecifics. Imagine two species of beetles, both of which secrete chemicals that are strongly deleterious to members of other species but not as dangerous for members of their own species. In this case which species survives in the container depends on the initial concentrations of the two species.

In the final case, one species is clearly better suited to the container than the other. It can then appropriate more resources than its rival and become the only species in the container.

These three cases are related to simple properties of the ecological niches of the two species. In the first case the container is considered to contain regions from two ecological niches. In the second case the container is in the intersection between two niches, and in the third case the container is within the niche of the victorious species.

It is possible also to construct equation systems involving many species: predators and prey, combinations of predators and competitors, and so on. It is also possible to add realistic details such as statistical variation, age structure, and genetic differences to the mathematical models, which generates more complicated conclusions. These can appear very authoritative, if only because of their complex mathematical notation.[38]

There have been persistent problems in applying many of these theoretical constructs of practical situations. One is that no populations, except for those of some microorganisms, grow according to the logistic equation of sigmoid growth used for the starting point of many theoretical arguments about populations. This S-shaped growth curve assumes, among other things, that all the individuals in a population are identically likely to die or give birth, that there are no time lags in physiological responses to changes in food supply or environmental quality, and that the physiological properties of newborns are independent of the properties of their mothers.[39] Despite these unrealistic assumptions, it still has some defenders,[40] who base their argument for its use on its extreme simplicity.*

Often simplifying assumptions are needed to permit the mathematical analysis. This introduces enough doubt to permit

---

* The basic logistic equation is $\frac{dN}{dt} = rN(\frac{K-N}{K})$.
N is population size, K is "equilibrium" level and r is instantaneous increase rate.

other advisers—and, in serious cases, attorneys—to deny the models' conclusions.

Nevertheless, mathematical models have suggested field and laboratory experiments. There are laboratory population studies, in which populations of two species were actually placed in a container, provided with renewable resources, and permitted to interact. Investigations of larger (visible to the naked eye) organisms required long periods of observation and repeated tedious counting. Single experiments could take longer than a year.[41]

Current studies of laboratory population dynamics mainly use microorganisms and are primarily of genetic and evolutionary significance. Even in these studies there are significant ecological differences between the results, depending on properties of the containers. Studies using liquids in test tubes give different results than studies in which liquid nutrient medium flows through the container, and these are different from results with bacteria growing on gelatin surfaces.[42]

An apparently simple single container can be divided between two species in interesting ways. For example, when populations of a moth and of a beetle were placed in a container with either intact wheat grains or flour, the moths were eliminated. Both species survived in cracked wheat. Broken bits of glass capillary tubing added to flour also permitted the two species to persist. Unless they were sheltered inside the cracks in the wheat grains, or in the glass tubes, the moth larvae were being eaten by the beetles.[43] This demonstrated that spatial complexity can modify the relation between competing populations.

As populations become more crowded there are changes in birth and death rates. These changes are always accompanied by other physiological, anatomical, and behavioral changes.

There is abundant evidence of psychological and physiological change caused by individual social history in mammals.

While the effects of past history may be most dramatic in mammals, they have been found in essentially all animals and plants. Crowded trees show different shapes than those grown in uncrowded situations.

The levels of crowding in laboratory population experiments are very much in excess of what the organisms normally tolerate. There is a general tendency of animals to escape from these containers in any way possible, the way cows or pigs would scatter if the fence around a feedlot were broken.

In one study it was possible to compare responses of house mice in closed and open spaces. When the confined mice became sufficiently numerous to empty their food trays they immediately developed reproductive and behavioral pathologies. Unconfined mice remained in a local area around their food trays until they could empty them after each feeding. They then immediately dispersed.[44]

Flour beetles when crowded move to the surface of their medium and show a strong tendency to fly.[45] This cannot be demonstrated in population vials but becomes apparent if the flour containing a crowded population of beetles is removed from the vials or the vials are unstoppered (personal observation). Dispersal when crowded is also found with other grain pests.[46]

Even hydra, which have no locomotory organs, respond to crowding and food shortage by floating free of their substrate, which removes them from a locally crowded situation.[47]

Laboratory experiments and theoretical models cannot adequately imitate nature.[48] However, an experiment can be designed to consider a wider range of conditions than those that occur in nature. In that sense nature becomes a subset of the experimental world.[49]

In some cases there is a clear logical transition between theory, laboratory experiment, and field experimentation. Theoretical analysis showed that predators might stabilize coexistence of two competing species by reducing their abundance so that all the organisms were getting enough resources to maintain R at some level greater than 1.[50]

I showed in laboratory containers that brown and green species of hydra would coexist if they were being fished out of the containers at a high rate. The green hydra have symbiotic algae that feed their hosts by their photosynthesis. This meant that the green hydra could overgrow and eliminate the brown ones. When the light was turned off, the two coexisted. Either predation or low illumination, a biological or a physical change, permitted coexistence.[51]

In a field experiment, Paine found that heavy predation of starfish on mussels prevented the mussels from eliminating their competitors.[52]

Theory proposed a possibility, laboratory experiments demonstrated its reality, and the field experiment demonstrated its occurrence in nature. The theory and the laboratory and field experiments demonstrated that under some circumstances a predator might enhance the number of surviving species.

But of course there are many examples in which predators wipe out or seriously reduce their prey. For example, feral house cats seem to now be destroying native species in Australia and Hawaii.[53]

In short, there is no simple theory of species diversity. The continued existence of any species depends on a broad series of requirements, different for each species.

I certainly am not against the preservation of species diversity. However, I cannot understand how the study of diversity in the abstract will lead to intellectual or practical advantages commensurate with the money, time, energy, and propaganda with which it has been surrounded. The Hans Christian Andersen story of the emperor's new clothes comes persistently to mind.

SPECIES EXTINCTION

Large predators tend to be rare wherever there are people about to record their rarity.[54] When the people leave they become less rare, as in the case of the wolves that have returned to the Italian Apennines now that the villages are in decline. Recently attempts have been made to reintroduce wolves in areas from which they have been eliminated; farmers protest. Bears have appeared in suburban neighborhoods in New Jersey, and some kind of balance must occur between concerns for the bears and concerns about the bears.

If an area is low in species diversity, it is either because extinction of local populations has occurred or because the area cannot hold more species.

Extinction is generally a bad thing, but that assertion requires defense. Who wants mosquitoes? There are five basic reasons to preserve species and populations from extinction. I will present them in order of what I consider descending significance. This is my personal order. Different ordering is possible.

First, I see each kind of organism as a masterpiece left to us from the past. Every species on earth has literally billions of years history behind it. Its ancestors lived through one crisis after another, day by day, year by year, and millennium by millennium. They have effectively taken advantage of all the good times and somehow survived the bad. They have changed in response to some pressures, hidden from others. I believe that it is horrible that the history of any species should be brought to an end by shortsighted or careless human activity or inactivity in my lifetime. This is my very personal belief. I hope others agree with me.

The second reason is a more practical one. There is much to be learned by careful study of any species, and so few species have been carefully studied that the extinction of a species, particularly an unstudied species, is intellectually equivalent to burning a book before it has ever been read. There is no guarantee as to what would be learned by the study, but the world would be a poorer place if the species were not there to be studied.*

The third reason is that wild organisms of many kinds produce saleable commodities. On a very practical level species extinction is generally bad for medicine and for other business. The spectrum of herbal medicines in the health food shop of your local mall will demonstrate how many and how curious are the virtues attributed to relatively rare plants. Sometimes these are used without proper clinical testing, but their role in folk medicine is of sufficiently long standing that proper testing ought to be done and the plants ought to be around for that purpose if no other.

Important medications, ranging from the quinine of the sixteenth century to the modern medicine for breast cancer produced from yew tree bark, have been found in wild plants. We don't know what medications and other useful products will be found in the future. Judging from the frequency of past discoveries, we can anticipate that dozens and perhaps hundreds of new ones will be found in the next century if the species that produce them survive.

A weakness of this argument for species preservation is that once a new biochemical has been found, studied, and synthesized, there might be no perceived reason to preserve the species from which it originally came. I do not believe that even a species that clearly has no commercial value and is not particularly lovable can be legitimately extinguished.

A fourth argument is that different people use a broad diversity of species, some of which are relatively rare and not subject to cultivation. The field of ethnobotany is now cataloguing

---

* This argument is due to Tom Lovejoy.

these organisms and their uses. Survival of these species may aid the cultural survival of the people that use them.

The fifth reason that has been proposed in favor of species preservation and biodiversity I find ingenious and fascinating but transparently specious and therefore dangerous both to the future of ecology and the future of the natural world, which we have accepted as our object of study. This is the "rivet in the airplane" argument.[55] It is a story that begins with elementary fact, then blossoms into a full-fledged parable and emerges with a reification, a creation of a mythical reality.[56]

First facts: It is a fact that natural communities have ten to ten thousand species in them. (That is, there are from ten to ten thousand species that can be retrieved from exhaustive sampling of a reasonably small area.) If a particular species is found in a community, it obviously has some role in the community. Should that species be eliminated, whatever the role of that species had been, it is no longer precisely filled, although various competitors can and do take over parts of the role.

Now the parable. Consider an ecological community as an airplane.[57] An airplane has many interacting parts. Imagine a passenger who finds that one screw in the cabin is loose and removes it. It will probably make no difference. But now someone else comes by, finds another loose screw, and pockets it. While a fair number of screws can be removed without evident harm, if a sufficient number of screws have been removed, the removal of one more screw will result in the plane crashing.

In the same way, states the parable, removal of one species from a forest may do no visible harm, but if enough species have been removed, the loss of one more population results in collapse of the forest community.

This is a beautiful and vivid image. The parable asserts that loss or addition of critical species will send reverberating waves of change through a community. Sometimes this appears to be valid. Is it universally true? In what sense is there really an airplane?

What usually happens when a species is eliminated from some region? Its parasites will also be eliminated. Its predators will be hungrier. Its prey will perhaps become more abundant, and in some cases there may be other effects. Grass may grow taller, sheltering more mice, or some tree may be unable to germinate its seed.

Calvaria major are large seeded trees, also known as monkey puzzle trees. Most known specimens in nature are very old. It was suggested that the large seeds of *Calvaria major* trees on Mauritius could not germinate unless they had been swallowed by a dodo and abraded in the dodo's gizzard, and since the last dodo was eaten by Dutch sailors around two hundred years ago, there have been no young *Calvaria major* trees produced.[58] This would have been a fine example of pulling out one part and making another part of the community disappear. Unfortunately, Quammen has debunked this myth.[59] There are in fact young *Calvaria major* trees in the forest.

American chestnut trees comprised almost a third of the large trees in southeastern American forests in the early nineteenth century. A bark fungus apparently introduced on Chinese chestnut trees around 1904 resulted in the death of three and a half billion mature American chestnut trees within the next forty years. Many of the roots survived, and the chestnut trees still exist as shrubby shoots from these old roots. These die of the blight before they set seed, and are replaced by new shoots.* While large chestnut trees dominated the forest before the blight, their elimination did not cause the forest to disappear.

---

* Introduction of resistant varieties of chestnut, including some hybrids, to American forests is encouraging the increase in populations of wild turkeys, which are highly dependent on chestnuts.

As noted by Hairston and colleagues, the southeastern forest is as dense with trees as it has ever been.[60]

Davis, studying the northward expansion of forests after glacial retreat, noted that each individual species seemed to migrate at its own rate—there was no movement of the forest as a community marching together.[61] Whittaker noted that the distribution of trees on various environmental gradients was as individual trees, not as communities.[62] MacArthur in his classical work *Geographical Ecology* built a theoretical argument for loose connections being the optimal relation between species in a community.[63]

We have intuitions about what forests should look like. As noted by Botkin, these intuitions are very personal and largely esthetic.[64] Even the boundaries between communities may have the characteristics of fractals rather than those of smooth lines.[65]

We do know that losing a particular species from an ecosystem or adding a new species may have consequences for other species. For example, loss of elephants or alligators would be expected to seriously alter the world for other plants and animals—elephants rip up or knock down trees, warthogs feed among the newly exposed roots, and baboons lift the loose rocks in the exposed soil for scorpions and other tidbits. Alligators in the Florida Everglades dig holes that retain water even when rain is sparse, and tufts of willow trees and other vegetation come to surround these alligator holes. Similarly, beaver dams transform streams into marshes and thereby provide habitats for water-loving organisms.

To eliminate elephants or alligators or beavers would radically alter the ecosystems. But such large effects are perhaps less interesting than more subtle and specific effects that are harder to see. Termites in East Africa build mounds up to six meters high that are islands on the flat plains. This permits some plants to escape the annual fires that sweep the low grasslands.

Floating fronds of Sargasso weed in the mid-Atlantic are the home for fishes and crustacea that cannot survive in the open water. Even the highly toxic Portuguese man-of-war acts as the home for a small fish that swims freely among its tentacles and shares its food. Obviously the detailed accounts of how species interact are voluminous and fascinating. However, we cannot dwell too long on them.

Don't these examples simply confirm the rivet analogy? Not really. Each example relates to a few specific species. The rivet analogy makes two assumptions: first, that there is no way of telling which rivets are important, and second, that there is an airplane. But there is no airplane! There may be a pile of parts, each one closely connected to a few others, and each pile only loosely connected to several other piles, but general collapse does not occur. I am not even certain what general collapse would look like.

Although the elimination of one species may not always result in a wave of other disappearances, some species are certainly more critical than others. Also, although I oppose unfortunate metaphors, which I believe are damaging to science itself, I believe most strongly that the actual global extinction of any species is a real and irremediable lessening of the richness of the world. Our avoiding species extinction is tantamount to passing masterpieces on to coming generations.

ARE ALL INVASIVE SPECIES VILLAINS?

There is an enormous literature related to how many species occur in particular locations. There is massive concern about the dangers of invasive species.[66] There even was a presidential edict on invasive species.[67]

It is often assumed that coexisting populations in some sense are a community, so alterations in one species will ramify through the others. Recently there has been a weakening of the basic notion of a natural community as a unit in any serious sense.

Invasive species are species that evolved somewhere else and then either moved to, were brought deliberately into, or arrived by accident in a new location, where they established populations. This is the inverse of the image of removing screws. What happens if new parts are added?

Invasions have recurred throughout the history of life. Animals have been transported on floating logs with seeds tangled into their pelts and in their guts and with parasites in their blood. Coconuts can float for a long time and germinate after they are stranded on a strange beach. Insects swept up in storms and birds blown off course by unusual winds may all find themselves in new places.

Given the millennia of exchanges, how has geographic heterogeneity persisted?

Organisms may be unable to travel across the space between two locations. The Bitter Lake in the Suez Canal seems to have prevented the movements of marine organisms between the Red Sea and the Mediterranean.[68] Gatun Lake's very low salinity in the middle of the Panama Canal acts as a partial barrier preventing the passage of the extremely venomous Pacific sea snake into the Atlantic.[69] Or if invasions do not occur, it may be due to lack of opportunity. Large mammals cannot float across an ocean on logs or hide away in ballast tanks.

If transportation is easy between two locations—say, two sides of a river or two nearby islands—the kinds of organisms in the two locations may be very much the same. Absence of particular kinds of organisms from one of the locations must then be due to differences in environmental conditions.

If transport between two locations almost never occurs, then the plants and animals of the two locations evolve separately and are likely to be very different from each other. Many South Seas islands were for most of their history completely lacking in mammals other than bats. They were also short on amphibians and even depauperate in birds. The peculiarities of these islands impressed early European explorers and strongly influenced the development of the theory of evolution by Darwin and Wallace.

Immigration is not easy. Most often organisms arriving in a strange place die. They are unable to find their niche in the new conditions. The reproductive process and very young offspring are sensitive to abnormal environments, so even if adults are transported and live in a new place, they are not likely to reproduce and establish a population. Most transplantations to new locations do not take. At least four attempts were made to introduce rabbits to Australia before one of them succeeded.[70]

Species that survive the migration process and actually succeed in developing new populations are referred to as *invasive species*. This has a generally negative connotation. It suggests problems analogous to those of human immigrants.

If invaders do manage to start a population, repercussions affecting the native species are to be expected. There are a sufficient number of unhappy cases known so that ecologists do not advocate the mixing of flora and fauna from different areas. Bad examples include tree snakes on Guam, walking catfish in Florida, tilapia and Nile catfish in Rift Valley lakes, kudzu in the southeastern United States, and cheat grass in the western United States. Even if deleterious effects have not occurred, invasion involves change, and ecologists tend to oppose change.

A famous invader is the Japanese beetle, a very beautiful animal in shades of green, brown, and gold. It is said to have arrived with the Japanese cherry trees planted in Washington.

As a child, I recall rosebushes stripped of their leaves by the Japanese beetle. In recent years I have the impression that the beetle has been much less destructive.

Accidental and deliberate transfers by people are major sources of invaders. These have been occurring ever since the first human paddled a log across a river, and they continue in the luggage compartments of jet planes. When the great canoes of the Maori came to New Zealand a millennium ago they carried the first yams, coconuts, rats, and dogs.[71]

There are domesticated species of fruits, vegetables, chickens, and cattle from around the world that we all enjoy. English sheep graze in Australia and New Zealand. Near Eastern wheat makes the staff of life. Tomatoes from South America are basic to the cuisine of Italy. The potatoes of the Andes are a worldwide staple. Sugar cane and rice, with the attendant slavery, remade the social and ecological structures of the New World tropics and subtropics.

Occasionally domestic animals and plants from Europe were simply turned out to fend for themselves in new places. Captain Cook liberated pigs and goats on Pacific islands that previously had no mammals other than bats, rats, and dogs. This was done with the humane intention of providing food for shipwrecked British sailors. If no pigs and goats had been made available, these shipwrecked sailors would have had to survive on fish, fruits, vegetables, crabs, lobsters, mollusks, and the quite edible Polynesian rats—not respectable foods for Englishmen.

Often plants and animals were introduced that were not intended to be of any practical importance but reminded the human immigrants of home. In the late nineteenth century a section of New York City's elegant new Central Park was set aside as a Shakespeare garden, intended to grow every plant mentioned in Shakespeare's plays. Blackberries, house sparrows, rabbits, blackbirds, starlings, and many more were given the opportunity to grow in America and Oceania. House cats in Australia and Hawaii are now in the process of eliminating or severely reducing native species.[72]

I am certain that attempts were made to establish the English robin in America, but I find no mention of them in scientific reports. I assume this is because they did not succeed. However, the nostalgic value of robins was great enough that the name was imported to America and applied to an American bird. The English robin is a small, nervous, and pugnacious bird with a red breast. The American robin is a good-sized thrush with a rust-colored breast that is very similar to the English blackbird.

I know some relatively inconspicuous invasive species that I enjoy very much. For example, I am fond of a bright orange-red raspberry from Korea that escaped from an arboretum and now provides refreshment to children along the back roads of Long Island. Among my favorites are two small aquatic animals that have traveled from Southeast Asia to the New World with a stopover at the Kew Botanical Gardens near London. One is the freshwater jellyfish, *Craspedacusta,* which escaped in Philadelphia around a century ago.[73] These freshwater jellyfish are extremely sporadic in their occurrence. I collected some in Gatun Lake in the Panama Canal.

A terrestrial flatworm of the genus *Bipalium* also passed through Kew Gardens. It is now in the process of invading America and is spreading throughout the warmer eastern states.[74] Since it is such a strange animal, I give myself the pleasure of describing it in some detail.

It is shaped like a planarian, but it can reach a length of twelve inches and has a front end shaped like that of a hammerhead shark. Its body is longitudinally striped in yellow and brown. Like all planarian-type flatworms, its mouth is at the center of its body.

I have fed it moderate-sized earthworms. It approaches the

worm with cobralike movements of its front end. The earthworms do not move once they have been touched. It then glides over the body of the earthworm and when its midpoint is directly over the earthworm its stomach everts like a white flower and surrounds the earthworm. The digestive process occurs in the everted stomach, and after around a half hour the flatworm crawls on, leaving a small dab of wet mud where the earthworm had been.

Bipalium requires high humidity but is damaged by liquid water. I used to be able to find it in the palm room of the University of Michigan Greenhouse, lying between stacked bricks. Liquid water seeped into the bricks, leaving behind pockets of damp air. Since the widespread use of pesticides in greenhouses I have been unable to find it at all.

Others have found it frequently in nature. Ogren has noted that it

> is an aggressive predator on earthworms. The spread and ecological impact of this flatworm will be determined in part by its interactions with potential predators and prey. In laboratory trials, we tested [its] ability . . . to prey upon earthworms of different species and sizes. We also tested the predatory responses of six salamander species and two snake species to the flatworms. It . . . attacked and ate members of all earthworm species offered and attacked earthworms over 100 times [its] mass. However, flatworm predatory success was related to the relative size of the prey. The largest prey eaten in our study was 12.1 times the mass of the flatworm that killed it. When attacking, it often used a previously undescribed behavior of capping the anterior end of the earthworm, causing subdued escape behavior. None of the amphibians and reptiles, tested as predators, treated it as a regular prey item. Only

> a few salamanders (2 percent) struck and ate a flatworm, with most salamanders and all snakes showing little interest in the planarian. Salamanders that consumed flatworms showed no apparent long-term ill effects.[75]

I hope *Bipalium* doesn't become a serious nuisance, but for now I'm glad it's here.

Also, there are other happy invasions. Striped bass were brought from the Atlantic to the Pacific coast of North America. In California they are called rockfish and are a prime game and commercial fish. Different species of oyster are now available at many locations. Gardens are full of exotic flowers. Of course there are gardens noxiously overgrown with things such as kudzu, but on the whole invasive species have enhanced rather than reduced species diversity worldwide.

Once populations of organisms are established, with the exception of very large and conspicuous animals, it is essentially impossible to eradicate them. The advance of the West Nile virus in the American Northeast is an example of a wealthy society exerting strong efforts against a mosquito without notable success. Dandelions, water hyacinths, zebra mussels, and green crabs are highly noticeable and very recent examples of successful and damaging invaders.

How can invasions occur if ecosystems are "packed full" of species, and what does "packed" mean in this context?[76]

The purple loosestrife plant is considered an invasive species. It is sometimes found in dense stands that keep out other species, but other studies suggest that purple loosestrife has no apparent effect on native plant species and may benefit native insect diversity.[77]

Apparently some invasive species fit into niches that were in some sense empty. Ruiz and Fofonoff estimated that more than 90 percent of alien species in estuaries have made no discernible impact on the species diversity or species abundance distribution of the estuary.[78] Levine and D'Antonio report a consistent positive relation between exotic species abundance and resident species diversity.[79]

Recently it has been shown that some species that meet the definition of *invasive* have settled into their new home and now have a broad range of ecological connections, such as might be expected of a long-term resident. Attempts to extirpate them may prove more dangerous to other resident species than leaving them alone.[80]

It is difficult to close borders against exotic fruits, vegetables, and insects, no matter how dedicated the border guards.* I believe that ultimately the fight against invasive species, despite enormous effort and a flood of scientific papers, will end in defeat. How much will it matter?

There will be changes—cases of native populations being extinguished, plant cover changing over large areas, diseases breaking out in new places. Some agricultural practices will change, but the likelihood of large-scale disaster is low, and in any case not much can be done to prevent the invasions.

Mark Sagoff, an interesting philosopher of science, has suggested that the invasions may not matter in any important way.[81] I am reluctant to agree with that, but my reluctance is related to my general preference for unchanging ecological situations. I am unable to strongly refute Sagoff.

There is a nightmare image of the world filling completely with dandelions, kudzu, and rabbits. It isn't realistic, and it will

---

* When we arrived in Kennedy Airport on a recent transatlantic flight, my wife had forgotten that she had two oranges in her purse. She would have been guilty of violating quarantine laws but for a charming border guard that sniffed out her transgression. The guard was a little beagle in a green sweater that sat in front of us and wagged its tail until a human guard arrived.

not happen. Rosenzweig suggests: "The breakdown of isolating barriers between biogeographical provinces will not have much effect on species diversity. In the short term, it will reduce global diversity but increase local diversity. . . . The considerable damage [some] exotic species have been known to do comes primarily from direct effects of particular introductions."[82]

Sometimes the invaders may be inconspicuous, but because we know the relevant history we are reasonably sure that elimination of one inconspicuous species will cause massive ecological changes. For example, the noxious plant Klamath weed was introduced into California, eliminating thousands of acres of good pastureland. In a masterpiece of ecological manipulation a species of beetle was found that lived on Klamath weed. When the beetles were liberated in quantity, more than a half century ago, the Klamath weed was brought under control, now existing only in small patches. The beetles are generally rare except when the Klamath weed patches become sufficiently large. Then the beetles congregate on the weeds and eliminate the patch. If the history of the situation were not known, we would not realize from collections or casual observations that the beetle is what is keeping the Klamath weed from overrunning the pasture. How many other relations exist in which an inconspicuous organism holds the key to maintaining an ecosystem?

Relatively common are inverse examples, in which a few seeds or a few specimens are introduced into an ecosystem and drastically alter it. When that happens we know that could we but eliminate the invader, there would be another drastic change in the ecosystem. Sometimes this is obvious. Less than fifty rabbits were introduced into Australia. They multiplied into millions and drastically changed the native ecosystem. If it were possible to eliminate all the rabbits, this would also drastically change the ecosystem—but it would not revert to what it was before. If we are concerned only with one species, we are talking about a ranching or farming situation. Other species must be considered in their relation to the one interesting species. They are helpful, or they are necessary, or they are weeds, or they are perhaps dangerous in some way as predators, competitors, or reservoirs of infection.

Recently several species of predatory birds were endangered by the widespread use of the insecticide DDT, which helped stop the spread of disease in refugee camps after the Second World War. One of the effects of DDT was interference with the birds' endocrine system, which altered calcium metabolism and therefore the strength of eggshells.[83] The eggs of large birds were breaking when the mother brooded them. Stopping the use of DDT has permitted recuperation of several species.[84] The presence of DDT is not the only recent environmental change that alters eggshell thickness, but nevertheless the elimination of DDT is a successful example of finding a subtle pollutant with serious and obviously real effects.[85]

3

Two Major Current Problems

Global Warming and Endangered Species

We have considered great global aspects of ecology, such as life in the oceans, and small aspects, such as how individual organisms survive and maintain themselves. We also have seen how populations grow and species are defined, why they are important, and how individual organisms, populations, and species may interact. I have assumed that most ecological change is dangerous or at least inconvenient and unaesthetic.

None of this is controversial. Human history, until recently, simply rolled on without seriously worrying about ecological changes. Humans are now numerous enough, and each of us is making sufficiently heavy demands on our environment, that we are, in the aggregate, a global force. We will make serious, permanent, and probably deleterious changes to the world even if we are extremely careful.

Realistically, we must deal with the sullied nature that surrounds us in such a way as to avoid excessive further damage. Ecology is now a matter of urgency, far from idyllic descriptions of Thoreau and of little boys wading in tide pools.

To reject the worth of nonpristine nature is to condemn all of humanity to depressing nostalgia for what no human has ever seen. Granted that a modern city and its surrounding suburbs seem more unnatural than a farming community, which in turn seems less natural than a hunters' camp surrounded by forest. It is a matter of degree. To a careful observer on foot there is still a great deal of nature visible even in a housing development.

How one treats nature depends strongly on knowledge of facts, but also on legal and political circumstances and ultimately on deep personal beliefs. The larger the scale of a problem, the more difficult it is to activate possible solutions, and the more strident the disputes.

There are now occurring relatively rapid, profoundly unpleasant, and far-reaching changes. Suggested mechanisms for ameliorating these changes involve massive expense, large-scale social reorganization, and unusual kinds of international accord.

WHAT CAN BE DONE ABOUT GLOBAL WARMING?

Up to now I have carefully avoided sensationalism and exaggeration. If anything I may have understated some issues. While I noted global warming as a merely curious fact in an ecology text I wrote forty years ago, it is now apparent that global warming is actually of enormous, earth-shaking importance. Excellent summaries of data and authoritative predictions demonstrating major consequences of global warming are available on various websites that will probably be updated while this book grows old. I will therefore not attempt to duplicate the information in them. (During October 2002 an excellent site was at www.ipcc.ch/.) Among the consequences are massive climate changes, new areas of both flooding or drought, species and population extinctions, human starvation due to loss of crops, economic disruption and changes in patterns of diseases of all kinds resulting from alterations in vector ecology.

Science has established the importance and urgency of the issue of global warming.[1] Clearly, solutions to the problems associated with this issue require scientific, political, social, and ultimately moral decisions made at every level, on an unprecedented scale. The scale of solutions must go beyond national borders.

There have been several international conferences attended by hundreds of scientific and political representatives. They have attempted to determine what each nation must do to help stop global warming. Some of what transpired at these conferences can be found on the web using the locations as key words. (Try "Kyoto" and "Rio de Janeiro"). There has never been such a focus of large scale concern on ecological issues. Various treaties have been signed in which governments make promises of ecological reform.

There is widespread concern at what will happen if ameliorative action is not taken. But many suggested actions are deeply opposed by those who feel that their financial or political position is more endangered by the remedies than by the ecological change.

Most of the realistic actions that might be taken in confronting the problem are curiously undramatic. How I eat and dress, how I maintain and heat my house, how cities are designed, how governments (local, state, and federal) exert power, and how we interact to maintain international compliance with regulations all affect global warming.

Briefly reviewing basic facts: Carbon dioxide can be removed from the atmosphere by green plants making carbohydrates. Carbon dioxide can also move from the atmosphere by dissolving in water, making extremely weak solutions of mineral water. And in some circumstances carbon can precipitate as insoluble carbonates or can be buried as organic sediment.

Carbon dioxide is added to the air by respiration of organisms, burning of any fuel, weathering of carbonate rocks, and emerging from water solution. Also, volcanoes add to the air new carbon that has never been respired or used in photosynthesis.

That is the basic list of the mechanisms for additions and withdrawals of carbon dioxide from the atmosphere. There is no global mechanism or natural regulator to make the rate of carbon dioxide removal equal to its rate of production.

The concentration of carbon dioxide in the air will be greater the more fossil fuels are burned, the more vegetation is destroyed, and the more rocks are weathered. If more carbon is buried out of contact with the atmosphere, the less carbon dioxide will be in the atmosphere. The greater the concentration of carbon dioxide in the air, the greater the growth rate of plants and also the more carbon dioxide will dissolve into the sea surface. I omit the details and even the quantities because we have already discussed them in part and also because for practical purposes most of the details don't matter.

Carbon dioxide is not poisonous for people until the concentrations are really very high. Slight additions of carbon dioxide stimulate human breathing.*

Measurements over the past century-and-a-half have demonstrated that the concentration of carbon dioxide in air all over

---

* People have fainted from being too near vats containing fermenting beer mash, where the carbon dioxide concentration is enormous. Some attacks of hiccups can be cured by breathing into a paper bag.

the earth has increased by more than 30 percent. It is a substantial increase.

Neither volcanic eruptions nor wildfires have increased very much during this time period, so it is agreed that the increase in carbon dioxide is due to human activity—specifically to increased fossil fuel consumption and to deforestation.

Even with the concentration increase, carbon dioxide is still a minor component of the air, slightly more than three parts per hundred thousand. During geologic time it has been higher—perhaps several percent—and perhaps as low as only one or two parts in a hundred thousand. Why does carbon dioxide concentration so strongly alter the temperature of the earth?

An important fact is that carbon dioxide, like most supposedly transparent things, is not quite equally transparent to all wavelengths of radiation. This seriously influences the temperature of the earth.

The radiation energy from the sun impinging on the earth precisely equals the amount of heat energy the earth radiates back into space. If that were not so, the earth would have been incinerated. While the amount of radiant energy impinging on the earth and that leaving the earth by radiation into space are the same, the *color* of the energy is different. This assertion may require some explanation.

We see in one narrow window of the spectrum of radiation; we call this visible light. There are other radiation wavelengths we cannot see, but we know that beyond the red of the rainbow there is infrared—a color with a wavelength longer than we can see—and beyond that is heat radiation.

When we park a car in sunlight with its windows closed, even on a cold day for even a half hour, the interior of the car will be considerably warmer than the air. The internal temperature of the car will continue to increase for a while and then the increase will stop. The interior of the car is warmer because the glass of the windows and windshield is not equally transparent to all the wavelengths of light. The multicolored sunlight that enters the car warms the seats and furnishings. These warm objects radiate heat. The car windows are relatively opaque to infrared radiation. The car becomes warmer but does not become hot enough to burn or melt because the rate of radiation from the car increases very rapidly with temperature.* When the car is warm enough it radiates as much heat energy out as it gains in light energy coming in.

Like glass, carbon dioxide gas is relatively opaque to heat radiation and transparent to most sunlight. It is therefore a greenhouse gas. There are other greenhouse gases in the atmosphere, particularly water vapor. All of the greenhouse gases together act like the roof of a greenhouse or the windows of the closed car. The earth's temperature rises until the radiation into space of the earth's heat equals the radiation from the sun that reaches the earth.

In short, an increase in concentration of atmospheric carbon dioxide should be expected to increase the surface temperature of the earth. Reduction in the amount of carbon dioxide would result in a colder earth.†

Opaque glass in a car would have the advantage of preventing radiant energy from entering the car in the first place and the disadvantage that the driver couldn't see where he or she is going. Similarly, dust and smoke—from volcanic eruptions or forest fires, high clouds or massive explosions—can prevent

---

* Radiation rate increases as the fourth power of the absolute temperature.

† There is evidence that during at least two periods of particularly low carbon dioxide concentration the earth developed a solid ice cover. It was rescued from being a global snowball by the heat and carbon dioxide from volcanoes. D. E. Canfield and R. Raiswell, "The Evolution of the Sulfur Cycle," *American Journal of Science* 299 (1999): 697–723.

solar energy from reaching the earth's surface, but this carries a disadvantage in terms of low rates of photosynthesis. Dust in the air has increased during the last century, but not enough to stop the changes caused by carbon dioxide.

Over the past hundred years the earth's surface temperature has increased by approximately 0.50 to 1.25°C. If current conditions of greenhouse gas production persist, the Earth's surface temperature will increase by another 1.8 to 5.8 degrees C. by the year 2100. The rate of temperature change will be higher than any hundred-year period for the last 10,000 years. The burning of fossil fuels, the clearing of forests for agriculture, urban development, and highways are reasonable explanations for the shift in carbon dioxide.

Supplies of fossil fuels are finite. There is dispute about how much actually exists, but it is clear that we are burning these resources at a rate thousands of times faster than that at which they were created. Millions of years of carbon burial have been burned in two centuries of industrial society. If we simply wait until all the fossil fuels are used up, the rate of increase of carbon dioxide in the atmosphere would decline. Unfortunately, we must face the consequences of increasing temperatures in the several centuries until that occurs.

What difference does it make if the earth's temperature is just a little warmer? So far the difference is not enough to affect the market for summer clothing. We don't really know all the consequences of increased temperature, but we do know some of them.

A very conspicuous consequence is the melting of glaciers. Glaciers are masses of slowly flowing ice and snow that now contain around 10 percent of the earth's water. They cover most of Antarctica, Greenland, and islands in the North Polar Sea. There are also glaciers covering parts of Alaska and Canada, and smaller glaciers on mountain ranges throughout the world—even at the equator.

Most of the world's glaciers have been melting. The state of particular glaciers is determined by local weather conditions so that some glaciers may actually be increasing due to changes in snowfall patterns while the earth as a whole is warming. Projections of melting including Greenland, but not Antarctica. If they all melt completely, which is certainly possible and has happened in the past, the water they now contain will enter the sea and raise sea levels worldwide. The rise in sea levels from complete melting would be around one hundred meters, or three hundred feet. That is approximately the length of a football field. Around 70 percent of the people in the world live within one hundred meters of sea level. Obviously, complete melting would be a cataclysmic disaster.

However, even a small sea level rise is a serious matter, since seashores are the most densely populated parts of the world. In some places, such as Bangladesh, there are broad areas of heavy habitation at just around sea level. Even a slight rise in sea level will inundate these and will permit high tides and storm waves to travel much further inland then they now do.

Changes in air temperature will cause changes in wind patterns and weather, drying some areas, adding moisture to others, and changing the geographic limits of many kinds of agricultural and wild plants and animals, including disease-carrying insects. Again, arguments are possible, but only about the extent of the problem, not about its reality.

There will also presumably be changes in ocean surface currents, changing fisheries. In short, there is a set of changes of just the sort to worry ecologists.

What can be done?

There is no single realistic, simple, and cheap technical solution to the array of problems.

Simplistic proposals have been made to grow more forests and thereby remove carbon from the air. Aside from issues of where these forests are to be grown and who is to do the work, this scheme will not work for elementary biological reasons.

When a tree is growing it is in fact taking carbon out of the air, but once the tree reaches its full size its respiration returns as much carbon to the air as its photosynthetic activity removes. When it dies and the wood rots or is burned, all the sequestered carbon is returned to the air. The net change in atmospheric carbon dioxide produced by any tree in a forest from its birth to its disappearance is zero. Planting a forest of young trees will temporarily remove some carbon dioxide from the air, but only until the forest becomes mature and stops increasing in total wood content.

If trees are cut before they rot and either buried to preserve the wood or made into objects such as houses or furniture, their carbon will not return to the air. The number of houses or tables that would have to be built to lower the carbon dioxide content of the air is staggering and utterly impractical.

It has recently been suggested that if we could find a way to increase the productivity of phytoplankton so that more carbon would sink to the bottom of the sea, we might sequester carbon from the air. There are two problems with this. First, what can we add to the sea to enhance phytoplankton enough to make a difference? Second, phytoplankton growth in the surface waters of the sea may not result in very much carbon sinking to the bottom. Unlike a lake, organic detritus in the sea decomposes relatively near the surface and may never reach the bottom. In the absence of ballasting by bits of shell or inorganic particles, carbon-rich particles may dissolve before reaching the bottom.[2]

While the carbon dioxide concentration change cannot be reversed completely, there are partial solutions that will ameliorate some aspects of the problems to some degree. They are undramatic, but taken together they would slow or even stop further global warming. Some of them would also ameliorate other social, political, and even ethical problems. The goal is not to achieve paradise but to avoid further deterioration of the world we have.

Any reduction in the use of fossil fuels would help. The most obvious mechanism is to reduce human population size. This would alleviate to some degree all ecological problems. Unfortunately, population reduction is a political, social, philosophical, religious, and pharmacological hot potato.[3]

Among the social changes that would probably reduce birth rates and population size are improved education and political and economic rights for women, opening careers other than motherhood. General literacy and social security may help alleviate the fear that those who do not have an abundance of children will suffer for it in old age. More than fifty years ago I heard the suggestion that good books and bedside reading lamps would be strongly effective along with more conventional birth control.* At a very small level we each have a contribution to make here, but notice that the social changes tend to be those of moderately left-wing politics.

There are also simple technological changes focusing more directly on carbon. For example, finding alternative energy sources, reducing the need for fossil fuels, and modifying how we live our public lives can all be helpful.

Preferable sources of energy, which do not contribute to the greenhouse effect, include burning of agriculturally produced plant materials, and water, wind, and nuclear power. Again notice the politically charged introduction of nuclear power. Nuclear energy generation has well-known dangers of its own. The assumption here is that the technology of nuclear power

---

* Conversation with Margaret Mead and Evelyn Hutchinson in 1949.

plant construction and inspection, nuclear leak containment, and waste fuel disposal all must be improved.

Wholesale switching away from fossil fuel sources also generates problems. To use land for fuel production we must either clear and develop presently unused land or use land that would otherwise be used for food and fiber production. This in turn will have ramifications for human food supplies and for the rate of species extinctions due to habitat destruction. Burning either vegetation grown specifically as fuel or animal wastes contributes carbon dioxide to the air, but both mimic the turnover of carbon in current ecological systems, a process that does not produce carbon beyond the global capacity for photosynthetic oxygen production.

There are also small, practical modifications of our lives that will modify how much fuel we each use and in the aggregate make a global difference. What modifications are in order varies from place to place.

An average citizen of the United States consumes energy at ten times the rate of a Chinese citizen. A sensible goal is to bring that rate down. It has been suggested on the American side that any treaty that lowers per capita energy use by Americans without reducing the use by Chinese or other citizens of countries that are using very much less energy per capita is in some sense "unfair." This seems analogous to asserting that income tax must be paid by all who have income, regardless of how small that income might be.

Some technological changes seem free of this fairness argument. Smaller, more efficient cars for most domestic purposes can cut fuel consumption per trip. Public transportation can do even more. Redesign of household utilities for efficiency is occurring and can proceed farther.

In summer, less air-conditioning is required if we wear light clothing and live and work in well-insulated houses with awnings or sunshades. Lowered thermostats combined with attractive sweaters and good insulation can reduce fuel consumption in winter.

There are larger-scale practical mechanisms. City planning designed to reduce travel and better public transportation would help.

These changes are not painful. They require effort and some organization, and the talents of people ranging from clothing designers to city planners.

In poor countries efficient stove designs and public transportation would cut down on consumption of fossil fuels, but there are obvious reasons why the total fuel consumption per individual in the poorest countries must in all reason be increased to improve living standards.

Like many ecological situations, global warming cannot be easily reversed. There is no cheap, immediate, politically possible grandiose solution. Nevertheless, the problem must be dealt with, and it can be dealt with, in part, if the world's citizenry becomes aware of it and each person modifies his or her behavior in relatively painless small ways. There is no certainty that these various recommendations will work, but in themselves they are all reasonable and contributory to social justice and welfare. They make sense even if there were no global warming problem.

PROTECTING ENDANGERED SPECIES

The loss of species has been a concern throughout history. Laws protecting particular species are as old as history. The Bible prescribes how and when one may take wild birds. In medieval England, swans and deer were royal property, protected from plebeian attack by preservation laws. These laws were designed to preserve ownership of organisms and not to preserve the organisms for their own sake.

Modern ecology has taught the importance of organisms in and of themselves. While various regulatory statutes existed in America from the beginning, the first major federal Endangered Species Act was passed in 1973. As noted by the Ecological Society of America:

> With the support of the Nixon administration, Congress almost unanimously passes a completely rewritten Endangered Species Act. The new law distinguishes threatened from endangered species, allows listing of a species that is in danger in just part of its range, allows listing of plants and invertebrates, authorizes unlimited funds for species protection, and makes it illegal to kill, harm, or otherwise "take" a listed species. In effect, the law makes endangered species protection the highest priority of government.[4]

The Endangered Species Act was considered to be a triumph of massive pro-ecological popular sentiment aroused by such authors as Rachel Carson.[5]

In addition to federal legislation there is various local legislation regarding endangered species. All these laws are now confronting the legal problems of ownership.

In a rather loose way humanity as a whole owns, or acts as if it owns, the earth as a whole. Ownership is at the heart of the American social and economic system. Subsets of humanity own bits and pieces of the earth, down to the point that my wife and I own our house and garden.

Organisms have specific habitat requirements for their survival and reproduction. Destroying those requirements is as deleterious as killing. Later developments of the American laws on endangered species took into account protection of habitat as well as direct attacks on the organisms themselves. Declaring a species as endangered asserts that these organisms belong to the government, whoever may own the region they occupy.

Endangered species regulations can usually be enforced on public land. However, in America private citizens or businesses often have been permitted to use nominally public lands for hunting, fishing, logging, grazing, mining, and recreation under particular agreements. When endangered species are found on public lands and their well-being is diminished by the way the land has customarily been used, major conflicts can arise. For example, the northern spotted owl lives on public land that has been customarily subject to logging.

More severe problems enter when we consider how the law can protect organisms living on private land. At this point technical aspects of law become interesting.

The answers to the questions of who owns what and what it means to own property are not always obvious. Things that are assumed to be in infinite supply or that permit unrestricted access, such as sunsets, have no owners. The fact that ecological problems concerned with air, water, waste disposal facilities, and organisms, requiring legal restrictions implies that each of these is perceived as limited and therefore "owned." Disputes or damages may arise either over immediate ownership or over the consequences of misuse by owners.

In what sense can we own nature? The animals and plants in our garden are part of the value of our property, although not easily directly translatable into cash. There are trees, shrubs, grass, weeds, and herbs. We planted some of these, but some were here before we purchased the land and others grew without our help. The plants that I planted are my property, to cut down, burn, or sell as I choose. However, the government of my village forbids me from removing so many trees, bushes, and other plants as to leave no cover for small animals. The village law doesn't care who planted the bushes.

Rabbits, squirrels, many sorts of birds (including a magnificent family of pheasants), a fox, garbage-hunting raccoons, and my neighbor's dog share the land with us.* Are they mine to do with as I please?

By general consensus I am permitted to kill rabbits any way I can. Killing pheasants, despite their edibility, would be a different matter. The pheasants wander beyond the narrow limits of my plot. If I killed a pheasant, even on my property, neighbors would complain that this beautiful bird was one of the amenities of our village, that I have taken it away from them, and that I must somehow compensate them for killing it.

I have occasionally seen on one of my trees a kestrel, the smallest of American hawks but with all the fierce dignity of its larger relatives. Perhaps it is mine while it sits on my tree and then becomes my neighbor's when it moves to a tree one hundred feet away?

In short, my property is mine in a very limited sense. I have essentially absolute rights of possession to anything on my property that no one else wants or needs. When needs or wants change, my degree of possession also changes. My possession is subject to the feelings of the community. In fact, I am expected to take care of some parts of my nominal property not because I want to, nor because it is profitable for me to do so, but rather as a responsibility of guardianship produced by being a nominal owner.

Swamps did not have value when I was a boy. They could be turned into garbage dumps, drained, filled, or ignored at the pleasure of the landowner. As a result of the rise of ecology, the

---

* Ancestors of the common pheasants were imported from Europe and turned out in America, but are now treated as natives.

designation *swamp* was changed to the designation *wetland,* and the owners of wetlands immediately lost some of their rights of possession. If one's intention was to leave the wetland alone anyway, this mattered very little, but if a landowner had decided to develop his or her land, either for commerce or for residence, then the new regulations protecting wetlands would in effect have taken part of the value of the land. Some of yesterday's perquisites of ownership had been taken away—whence the general notion of takings.

Twenty years ago common box turtles wandered across Suffolk County, eating in gardens, lawns, and uncultivated land. When I found them they were often eating wild strawberries, which on Long Island are beautifully red but dry and seedy and best eaten by turtles.

One day, around twenty years ago, I ran over a box turtle with the lawn mower and killed it. I realized then that box turtles were the wrong shape for lawn mowers. The high domed shell was just the height to be cut off like the top of a boiled egg. Eventually they were listed as an endangered species. Now it is up to me to make sure I am more careful with my lawn mower. My killing of a box turtle today could result in a heavy fine.

What if I didn't actually kill a box turtle but extended my neat lawn into the bushy corners, wiping out areas where the box turtles were safe from mowers and also wiping out the wild strawberry plants that had been so attractive to the turtles years ago? Then I would be taking habitat away from an endangered species and violating the Endangered Species Act.

What if I cut down trees on my property that some endangered species of bird used as a nesting site? I would, in fact, be taking a nesting site from the birds, and this act of taking reduces the probability of the species' survival by some amount. Under one interpretation of the Endangered Species Act I could be enjoined from cutting down my own tree or building on my own land if this constitutes a taking from the habitat of an endangered species.

The habitat of the Florida scrub jay is a special combination of sandy soil and xeric plants including wild rosemary. In the early 1990s word went out that Florida scrub jays were likely to be declared endangered species. Landowners immediately began bulldozing and chain-logging the scrub before the law went into effect. The land was good for growing grapefruit.

The problem of responsibility for an endangered species does not end at the borders of private property. Consider places where endangered species are found on public land adjacent to private land. Assume that the private landowner has prudently wiped the species in question off his or her land before anyone knows it is there, solving most of the problem. Imagine the surviving nests or burrows are on public land but the organisms wander onto the private land. In proper jargon, the private land is part of their *range*.

Is the landowner prevented from damaging them? Certainly, since the endangered, and therefore protected, status applies to the species, not to the real estate boundaries. What if the landowner chops down trees that had been used as perches or rubbing posts or whatever? This becomes ambiguous.

Most endangered-species laws during the early days of ecological consciousness merely prevented direct damage to endangered species. It quickly became apparent that rare species could be severely damaged or destroyed without being hunted or fished or poisoned.

Each species has habitat requirements—feeding places, nesting places, hiding places, and so on. The meaning of species protection was extended by officials of the federal Fish and Wildlife Service to protection of the habitat of an endangered species. The legal authority for this extension was not explicitly written into the law but was derived from the Fifth Amendment to the United States Constitution. This amendment is part of the Bill of Rights and is usually invoked as protection against self-incrimination. But the wording of the amendment proceeds beyond self-incrimination to say: "No person shall be . . . deprived of life, liberty, or property, without due process of law; nor shall private property be taken for public use, without just compensation." This section of the law is central to the maintenance of legal property rights of many kinds.

The argument of the Fish and Wildlife Service was that just as there are restrictions on activities that take away value from personal property, destruction of habitat takes benefits away from resident species. To build a tall factory building on the plot of land next door would destroy the beauty and, more importantly, the resale value of my land. It would constitute a taking from me, one that could perhaps be compensated for by a cash payment making up for the diminished value of my property.

In an analogous mode, any activities that decreased the likelihood of survival of an endangered species could be considered as a taking from the endangered species, which is the property of the people of the Untied States, whether or not these activities took place on a reserve or on private land.

This understanding was seen as a powerful weapon in the hands of conservation agencies, until an analogous argument was used by advocates for private development rights to essentially eliminate most species protection legislation.

If I concede that the state is acting within its rights to constrain my use of my property for the sake of wildlife, the Fifth Amendment requires that I receive fair compensation for rights to my property that are taken.

Takings are a special complication of the right of eminent domain. At its simplest, eminent domain involves a transfer of property from a citizen to a government and appropriate compensation from the government to the citizen.

Imagine that the county owned all the land surrounding my plot and decided to use the land as a garbage dump. Being embedded in garbage would materially lower the resale value of my plot. The Fifth Amendment would give me the right to sue for the fractional loss of value of my property. That is, *taking* refers not only to transfer of property but also to loss of value of property that has been caused by some action.

Currently there exists pending legislation that would require the federal government to pay a landowner in any case in which 10 percent or more of the fiscal value of the land is lost because of an endangered-species regulation. In effect this prevents any more species from being declared endangered by the federal government unless the appropriate agency has the money to defray costs to landowners.

The situation is in reasonably dramatic flux because there is a converse possibility based on the idea of nuisance. What if I use my property in such a way as to diminish the value of property belonging to others? I might make a garbage dump! Governmental authority could enjoin me from building the dump on the grounds that such use would be a public nuisance, taking value either from my neighbors or from the nation as a whole.

Now, let us say that instead of building a dump I killed a member of an endangered species—I had a box turtle feast, for example. It is possible to consider that killing the box turtles was also creating a nuisance, diminishing the quality of life of my neighbors or even of all the people of the United States. It would be hard to assign a dollar value to this diminution, but it might be done.

If I have lost resale value or business opportunities because of a government regulation and have not been compensated, I am suffering from a taking. However, it can and has been argued that refusal to abide by governmental regulation constituted a nuisance and therefore I am entitled to no compensation and may even be subject to penalty.

Law courts are filled with suits and countersuits describing nuisances and reversed nuisances. Out of the sound and fury the lawyers do reasonably well, and there is some clarification of the meaning of the law.

Lawsuits are settled in courtrooms in accord with common case law as distinct from law promulgated by Congress. Sagoff has made a survey of cases in takings law and comes to the conclusion that as of 1997 "takings jurisprudence has become the night in which all cows are black." All the "distinctions on which a substantive legal theory might rest have collapsed."[6] This curious situation has arisen because there are environmentalists who make claims for takings having occurred based on their definitions of what ecological systems require or consist of. Sometimes these definitions are curiously metaphysical or idiosyncratic, and often they are readily attacked by opposing lawyers.

For example, consider the criteria for ecological damage generated by the assertion of the famous ecologist Aldo Leopold that a "thing is right when it tends to preserve the integrity, stability, and beauty of the biotic community. It is wrong when it does otherwise." Or this: "Morally acceptable treatment of the environment is that which does not upset the integrity of the ecosystem as it is seen in a diversity of life forms existing in a dynamic and complex but stable interdependency."[7] (If that doesn't sound like nonsense to you, I have failed.)

In the absence of a clear set of ecological criteria, how can the legal tangle be simplified? In fact it cannot be. Each case must be decided on its own in the context of local political realities and continuing vigilance.

Taking and just compensation require precise legal interpretation, and these interpretations now stand as the main impediment to activation of the Endangered Species Act and perhaps to ecological regulation in general. This is not a matter of obfuscatory legalism but rather strikes at philosophical issues central to applied ecology. The ramifications of an analysis of the idea of takings are of enormous political, economic, and even philosophical importance.

Generally, in the past the difficulties and ambiguities of ownership have been ignored by ecologists. Notice that the Ecological Society of America has recently produced two major manifestos on the future of ecological research and why it is vital to the world that it be generously funded. The word *takings* does not occur in either of these.

Underlying the use of the idea of taking in the context of environmental legislation and jurisprudence lurks the question of what it means to possess nature.

Perhaps nature is its own possession, but that assertion solves nothing. We may claim we hear nature, but in fact nature has no voice of its own. This is not a new idea. It is stated with reasonable clarity in the Nineteenth Psalm, which states that while nature may declare the glory of God, there is neither voice nor hearing nor any sound other than through human intellect.*

The assumption that since some divinity owns the world, that divinity will take care of the world, so no restrictions of any sort need be placed on its inhabitants, seems patently absurd. This absurdity has actually been the policy of some governments and religions. But the dominant metaphysical position of most nations, peoples, and religions is that regardless of actual ownership in a metaphysically correct sense, we had best behave as if we, collectively, were the owners.

---

* My reading of the psalm is based on the Hebrew text. When the English translations were made sometimes words were added to make more sense in English. These words are printed in italics in most English-language Bibles. To find my reading, omit the italicized words when you read from an English-language Bible.

I attended a debate between Justice Antonin Scalia of the United States Supreme Court and Rabbi Adin Steinsalz, whom some regard as the greatest Talmudist since the Gaon of Vilna 250 years ago. The subject was something like this: Does natural law exist, or is all law created by political process?

The rabbi favored the existence of natural law. If there is natural law, then God, or perhaps nature, is its author. This would demonstrate that there are forces beyond politics that regulate our behavior.

Justice Scalia's arguments were strongly in favor of the idea that even if there were something that someone might call natural law, it would appear in any debate or court in the voice of a person and would have to be treated like any other statement. It would have to be tested by case law precedents. In this view we are closed to purely metaphysical solutions.

This is an extremely important conclusion. Faith in nature or any other faith is a private matter. It may be indispensable to private tranquility, but it is not legally relevant to any decisions whatsoever. Metaphysics can build desires but cannot be coercive if science is to survive. If science does not survive, we have no objectively testable source of information.

To find practical solutions to all but the simplest practical ecological problems requires that we begin with approximately valid information and then face the army of nonscientific issues that must also be included. The overall ecological situation of the world is sufficiently serious that we cannot wait for either perfect information, a perfectly reliable way to judge incomplete information, or a perfect way to settle disputes. Many decisions must be made now.

4

Applying Ecology

Experts, Pseudoexperts, and How to Tell Them Apart

All suggested solutions to practical ecological problems have some common properties.

There are many kinds of practical ecological problems. Clearly solutions suggested to protect rare species differ from those suggested to alleviate global warming or pollution. However, there are some common features to all good attempts at solutions. Each solution must take into account the full consequences of action or inaction, as far as possible. For example, plans for the use of river water for agriculture must not destroy the usefulness of the same river for shipping, swimming, or drinking. Good applied ecology takes into account all externalities.

The term *externalities* come from economics. It refers to those costs and consequences of a process or activity that are borne by society rather than by the people who benefit from the activity. For example, in the nineteenth century the loss of limbs or life by workers in farms, mines, or factories was usually an externality. The owner of a mine, mill, or farm was not responsible for the support of damaged workers or their families. It was understood from the beginning that employment had its dangers. It was assumed that society at large would somehow pay to support cripples and orphans.

Today maimed workers may be supported by government pensions or private charity, in which case the cost of their support is still an externality to the business that owned the facilities in which the workers were hurt. Increasingly, the costs of maintaining damaged workers are being *internalized,* so that the costs of the products reflect the costs of maintaining damaged workers. The costs of workers' compensation insurance are either deducted from profits or added to consumer price.

In a more ecological context, early manufacturers could freely pour their factory wastes into streams, making them more or less useless for other purposes, such as watering cattle or even irrigation. The privilege of contaminating the water supply could be understood as an externality of a moneymaking process—the state or the general public could take care of the pollution, permitting the manufacturer to sell products at a low price. If it is required that the manufacturer either not pollute or clean up the pollution, this would *internalize* the costs of pollution, raising the price of the product.

Higher product costs weaken the competitive position of a nonpolluting manufacturer compared to polluting competitors, local or foreign, that are free of these costs. To simultaneously preserve profitable industrial, mining, and agricultural activity while internalizing the costs of maintaining environmental quality is a serious challenge. Failure to solve it can result in either hazardous pollution levels or loss of jobs.

A first activity for applied ecology is to clarify the existence and cost of externalities and attempt to internalize these costs without damaging the competitive status of those that comply with regulation. How this is done will depend on local circumstances, social attitudes, and technical ingenuity.

If consumers refuse to purchase products that fail to meet certain standards, this can compensate for the competitive advantage of ignoring externalities. For example, the enhanced salability and purchase price of dolphin-safe tuna may eliminate one externality of tuna fishing.

Many years ago Garret Hardin presented the idea of the "tragedy of the commons."[1] This curiously simple example has gained popularity as an argument in favor of laissez-faire capitalism.

Hardin imagined a village with a commons—a piece of land that belonged to the entire village and on which each villager was permitted to graze his or her own animals. Hardin said that each villager would be expected to gladly care for his or her own animals but could not be expected to care for the commons itself—for example, manuring or weeding to improve the entire meadow's grass. Any money or effort put into the well-being of the commons would immediately be dissipated among all the villagers and not be returned to the one who had paid for the improvement. If the commons was replaced by individually owned grazing plots, each owner would have an interest in the quality of his or her grazing land, and the entire system would be in better condition. Hardin's story has been widely quoted as a demonstration that communal ownership of a resource necessarily results in its degradation.

Does this story constitute a real proof of anything? I do not think so. I can use the same story to demonstrate the superiority of pure communism. Imagine that not only the land of the commons but also the beasts in the grazing herd are communally owned, with villagers getting a share from the sale of the entire herd. Hardin and others have suggested that in such a situation no one has any incentive to work as hard as possible because the rest of the community would pick up the tasks of a slacker. Perhaps, but I can imagine ways of discomfiting slackers. Certainly the arguments of the "tragedy of the commons" are not hard-and-fast laws.

In addition to taking account of externalities, all ecological advice deals with future uncertainty in making present solutions. Frequently the occurrence of ecological problems can be anticipated, not prevented; however, we cannot be sure of the exact form they will take, nor can we wait until they are upon us to decide on action. This has been most clearly analyzed in the reports of the U.S. Congress's Office of Technology Assessment.[2] While global warming is inevitable, we do not know exactly how much the increase will be. Also, given patterns of wind and ocean currents, it is generally not possible to firmly predict how different spots on earth will be affected by global temperature change. We could ignore the problem, except that the warming is occurring now. How does one deal today with the prospect that water regimes will change within the next decades or century, although one cannot be sure if this will mean flood or drought in any given area?

The Office of Technology Assessment suggests that one should focus on seeking *no-regrets, flexible, robust* solutions. Each of the italicized terms embodies a serious theoretical construct. *No-regrets* responses are those that are simultaneously of value now and relevant to the possible future; if the future development does not conform to expectations precisely, it would still be of value to have undertaken the response. *Flexible* solutions are those that can be modified easily and can be subjected to midcourse corrections and fine tuning. *Robust* responses are those that would still be of value under the worst conceivable case. For example, in response to possible changes in rainfall patterns, modifications in the patterns of water use are preferable to dam construction, other things being equal. Regulation of which crops can be grown, how water is taxed or paid for, and how flowing water is used for waste disposal all can be quickly modified to conserve water, and should water become more available, the regulations can be relaxed. Contrast this with dam construction. Massive resources are permanently built into an inflexible structure. A dam may very well be the only solution under some possible futures, but unless these futures are absolutely inevitable, dam construction should be delayed.

Any managerial decision about any major ecological problem will benefit some people more than others, cost someone money, and inconvenience somebody. This may lead to quarrels. Quarrels may be settled by means ranging from an argument over a backyard fence to legal sanctions or lawsuits between states and on up to alteration of world trading treaties. At the extreme case, in which people give up on rational solutions, quarrels may lead to war.

Some of the most interesting controversies within ecology include:

- How much and what kind of fishing may be done on particular fish populations?
- How should rare species be preserved, and how much economic sacrifice should be made in order to save particular species?
- What are the relative merits of biological and chemical insect control, and how can optimal management decisions be enforced?

In addition to ecologists, participants in these decisions may include pesticide manufacturers, farmers, "green" political parties, scientific publicists, journalists, leaders of citizens' organizations, and many more.

The following discussion presupposes that the ecological

information involved is of the best quality available. Clearly, innocent nonsense about ecology is as dangerous as deliberate lies or errors.

Many practical ecological problems relate to shortages. Water and space are perhaps the simplest to deal with. Problems of water shortage, at their simplest, may concern farmers, fishermen, city managers, recreational boaters and swimmers, and rival governmental units. They also concern manufacturers that use water in streams as raw material, coolant, or even waste recipient.

There have been vituperative disputes about water fluoridation, the allocation of water between multiple users and multiple political entities, the effects of withdrawal of water from the Mississippi River system on shipping, and deep issues of the effect of water allotment on issues of war and peace. Often actual decisions are compromises, based as much or more on politics, propaganda, or even military power as they are on considerations of ecology.

For decisions about essentially all problems that actually are of practical importance, confrontations of opposing viewpoints will occur. These confrontations will involve "experts." Experts are claimants to special competence. They are supposed to provide helpful advice about solving particular practical problems. Some are more useful than others.

Science provides knowledge, which is only one major qualification for being a legitimate expert. Ideally, experts provide wisdom as well as knowledge. Science does not provide wisdom, which is a less straightforward concept. How to produce genetically modified crops is a technical question answered on the basis of knowledge. Whether or not to produce them is a choice involving wisdom.

An expert ideally has the capacity to make significant "correct" decisions about real problems, even if the problems do not conform to the usual rules for scientific questions. Many different paths to wisdom have been suggested. There is general agreement that knowledge somehow helps. Contemplation of erroneous facts and theories may result in dangerous illusions.

In the network of issues that surround large-scale environmental decisions it is very difficult for concerns about "details" such as the survival of rare species to carry weight. For example, most ecologists would agree that the rivers of Appalachia *must* be managed so as to avoid damaging endangered species of snails and fishes. But each of the users of a river can make a case for some particular focus that *must* be central. Who is competent to choose among central foci and how should they be empowered?

The general public asks sciences that are defined by their subject matter, including geology, chemistry, medicine, and ecology, to answer all questions about their subject. Geologists are asked all questions about rocks, physicians about human illness, and ecologists about organisms and their habitats. It is hard for scientists to remain silent when confronted with a question about their subject matter, even if there is not as much information as they would like.

Many ecologists fear that if they refuse to attempt a reply to some particular question, then answers provided by people with dangerously little understanding will be used instead. For example, if there is a dense population of mosquitoes in a residential neighborhood, ecologists will try to lower their numbers—perhaps by draining standing water, fixing screens, and suggesting insect-appropriate dress. Engineers or chemists confronted with the same mosquitoes often use lethal fogs of insecticide that will temporarily lower the mosquito population but also lower the populations of many other organisms.

Experts have a role whenever scientific knowledge, public policy, management, and financial issues come together.

Because the boundaries of practical problems need not conform to boundaries of knowledge, honest differences in scientific opinion and advice exist and are important. It is often extremely difficult to evaluate supposedly scientific assertions in areas beyond our own training. If facts and their implications were perfectly clear, there would be no need for experts. But in the real world, competent experts are valuable whenever serious decisions must be made.

Even the best of experts may be in error. Sometimes explanations turn out to be wrong. Being wrong occasionally is an integral part of advancing science. I don't advocate it, but I can excuse it. However, I cannot accept dangerous pseudoexperts. This raises the question of how we can evaluate the competence of someone who was hired because his or her competence is supposedly higher than ours in some particular field.

Recognizing scientific-sounding nonsense can sometimes be difficult, but here are a few guidelines. Evaluating experts is a big topic that requires its own expertise. I cannot develop it as it deserves, but omitting it entirely is a disservice.

All experts use words, diagrams, or mathematical models to present their conclusions. These can be used to clarify or to obscure. In deciding how much to trust an expert, simple considerations of language and how it is used are often helpful.

Watch out for opaque language and distrust the experts that use it. Opaque language is used when there is too little information to make a clear statement or when there is some agenda other than transmitting information. Slow, careful reading and attentive listening can help. Summaries and speed reading are dangerous if actual decisions are involved.

Sometimes opaque language can be recognized by peculiarities of vocabulary, sometimes by peculiarities of logic or expression, and sometimes by odd choices of information to transmit.

Pseudoexperts may suggest that the two sides of a controversy have attracted different kinds of people. If you can establish that you and your listener or reader are the same sort of people and your opponents are something foreign, it becomes easy to sell lies. This may be done by presenting caricatures of their opponents: "We are trustworthy. They are not."

Ecologists may be portrayed negatively as sandal-wearing, tofu-eating, overeducated tree huggers who are descended from effete recent immigrants without even a hint of manure on their shoes and who can't drink beer from long-necked bottles. On the other hand, and with equal lack of legitimacy, ecologists and their friends may be portrayed positively as sensitive, informal, outdoorsy descendants of woodsmen and pioneers.

Opponents of the desires of ecological organizations are sometimes portrayed as classic exploiters, perhaps with round bellies, fur-collared coats, and jutting cigars like the cartoon capitalists in the *New Masses* magazines of the 1930s, or as ignorant rednecks.

In New York State political candidates often emphasize that their opponents came from New York City, that well-known den of urban and foreign iniquity, rather than from the apple orchards and vineyards of upstate. Conversely, upstate candidates are not quite called "apple knockers" by residents of the city.

Anyone who defines a class of friends and a class of enemies is to be suspected of using opaque language.

Another standard method of obfuscation is to shift the emphasis of a problem. For example, Congress recently proposed taxes on cigarettes with the intention of limiting the ability of young people to buy them. The major tobacco companies purchased highly visible advertisements on TV and in newspapers. The ads completely ignored the motivation for the taxes and focused entirely on the fact that the taxes would be paid by the poor and ignorant purchasers of cigarettes. The companies claimed that once again these working people were to be victimized by a government whose only motive was to take their hard-earned money by regressive taxation.

If your conclusions claim to have been drawn from accepted wells of truth, it is relatively easy to build confidence as a prelude to deceit. In America at the moment, the most convincing sources of truth are science and religion, in that order. Religious language is not usually relevant in public debates. I suggest that we avoid it in our own arguments and distrust it in the arguments of others. If we must resort to religious or quasi-religious arguments, as I did in discussing the importance of preserving endangered species, they should be carefully labeled.

Another danger sign is a claim to arcane wisdom. Don't buy snake oil! Advertisements for the modern equivalents of snake oil may betray themselves by using words that were obviously manufactured to mimic technical jargon, such as *cellulite*.

Reference to "down-to-earth" knowledge as opposed to "high-flown" theories is also a danger sign. Consider the myriad television testimonials to the effectiveness of patent medicines and foot powders from supposed beneficiaries, all living in very small towns.

Almost anything touted as being the newly discovered complete solution to any major problem (ecological, political, or medical) is almost certainly flawed. Modest claims can be tentatively accepted. Becoming maddeningly sexually attractive or ensuring immense retirement income are major problems in this context. Keeping squirrels off a bird feeder is a minor problem.*

---

* There is actually a product that consists of ground hot pepper, which, when added to birdseed, really does repel squirrels and is not tasted by birds. I assure you it works. It also works beyond its claims. It repels raccoons. Its effectiveness against raccoons may be claimed in the future since my wife gave a testimonial to the manufacturer, who seemed quite grateful.

Sounding excessively certain is a dangerous signal. For example, a CNN news program makes the proud claim "No mathematics! No statistics! Just the facts!" An absence of statistics, or their misuse, is a danger sign. One of the most important ways of distinguishing truth from nonsense is precisely the use of statistics. I do not feel reassured when a leading news channel assures me that numbers are unnecessary.

Information about data is not too difficult to transmit. Recently the results of political polls have been presented with an explicit estimate of their probable error. This is important. Some measure of error is inherent in any scientific measure, and failing to mention it is either sloppy or malicious.

Statistics are evidence about believability. If we are asked to believe assertions without evidence, we are relying on either wisdom or on wizardry. We want the first and should by all means distrust the second. But even numbers and graphs can obscure truth.

Watch out for "wizards"—people with claims to secret sources of knowledge. Edison was the "Wizard of Menlo Park" and Carl Sagan the wizard of the stars. Blacksmiths were considered wizards by many peoples of central Africa. Faust and other doctors had wizard status in Europe. Even in modern America scientists and engineers may have reputations for wizardry—for example, the proverbially wizard "rocket scientists." Good experts disown wizardry.

Wizards carry paraphernalia. Pointed hats and magic wands do not carry authority outside of children's movies, but we have computers and mathematical symbolism. The fact that a calculation was made on a computer is not at all related to scientific validity. Watch out for paraphernalia and obscure symbols.

There are some more subtle signs of pseudoexpertise. Unfortunately, many of them involve deeper analysis of mathematics and statistics than I can present here. For many purposes, however, even a superficial look at mathematical and statistical argument can at least indicate potential dangers.

If an expert claims not to be making any assumptions, he or she is not to be trusted. If arguments are presented in the form of complicated-looking equations, find out what assumptions are being made. These should be listed clearly. If they are not available, that is a danger signal.

If the assumptions are obviously incorrect, you need not read or hear the rest of the material to know that it is nonsense, regardless of how complicated the mathematics or the words or the graphs. Astrologers use very sophisticated computerized databases derived from astronomy and from the unproven assumption that stars actually do control the events in human lives. The result is no more valid, however complex the computations. Complex structures built on nonsense remain nonsense and are not worth the time it takes to refute them.

Of course, you must be careful to ascertain that what seems nonsensical to you really is nonsense. Unfortunately, to make this determination you may have to consult with an expert that you trust on other grounds.

In general, in any discussion of environmental problems not only are experts necessary, but you should have your own expert, just as you should have your own lawyer when you are buying a house, getting divorced, or preparing to die.

As in legal affairs, it is advisable to distrust your opponent's experts. Also, bear in mind that in almost every kind of environmental dispute, lawyers are at least as important as scientists. However, any lawyer's argument using incompetent science is weak and unconvincing. This can be demonstrated by persuasive experts.

Experts in general are paid for their services. They may be willing to adjust their arguments to simultaneously increase the satisfaction of their clients and enhance their own reputations. This is not necessarily illegitimate, but it does bear watching.

Many ecological quarrels are between curiously unmatched opponents—perhaps an oil company against a group of local fishermen. The chances are that the highest-quality—or at least the most expensive—experts will be working for the oil company. On the other hand, any lawyer is strongly motivated by the fact that defeating a powerful opponent will materially enhance his or her status and probable future income.

It is becoming usual for organizations with large financial stakes in the outcome of a dispute to hire experts by the dozens, partly to support their position and partly to hire away possible opposing opinion. Management decisions are generally not made on the authority of one ecologist alone, except in cases of strongly charismatic or dominating personalities (names on request). Usually the ecological expert supplies advice to some person or group with greater authority, who may or may not act on that advice.

Experts also offer general advice, outside of the context of specific quarrels or confrontations, just as doctors advise against smoking, basing their advice on sound science. Nothing prevents pseudoexperts from offering general advice. Again, wisdom is required to avoid nonsense.

Most pseudoexperts are not deliberately evil and are only trying to make a living. Nevertheless, they must be watched carefully. They are potentially dangerous.

THE IMPORTANCE OF BEING NATURAL AND VEGETARIAN

From the standpoint of personal health, I do not know of any strong evidence of the superiority of restricting my diet to fruits and vegetables.* An omnivorous diet seems to work well for

---

* Opponents to meat eating on moral grounds have included George Bernard Shaw, Mahatma Gandhi, and Adolf Hitler, as well as innumerable adolescent girls.

most people. There is also an anatomical argument that an omnivorous diet is natural for people: Human teeth are capable of both the slicing used in chewing meat and the crushing used in chewing vegetables. All animals whose teeth have these properties are omnivorous. These include pigs, bears, raccoons, and opossums, among others.

Although the shape of our teeth seems to indicate that omnivory has been our evolutionary situation, we are not so limited by our evolution that we cannot choose to limit our diets to plant products.

Vegetarians claim something like the following: "From an ecological standpoint, eating vegetables is generally less demanding on the environment than eating meat. Reducing meat production by just 10 percent in the United States would free enough grain to feed tens of millions of people. A pound of wheat can be grown with 60 pounds of water, whereas a pound of meat requires 2,500 to 6,000 pounds of water to produce. Energy-intensive U.S. factory farms generated 1.4 billion tons of animal waste in 1996, which, the Environmental Protection Agency reports, pollutes American waterways more than all other industrial sources combined."[3] I cannot dispute these claims.

Also, people may have strong moral feelings about killing all animals, but usually not about picking fruits and vegetables.

Benjamin Franklin describes how he was convinced he should be a vegetarian on moral grounds until he saw codfish brought on board a sailboat. He wrote, "When the Fish were opened I saw smaller Fish taken out of their Stomachs—Then, thought I, if you eat one another, I don't see why I mayen't eat you."[4]

I cannot imagine any discovery by ecologists that would seriously alter the moral convictions of vegetarians, one way or the other.

It has also been suggested that there are deep biological differences between plants raised on fertilizers from plant and animal manures and those raised on chemical fertilizers. The chemical fertilizers are seen as "unnatural" and therefore unhealthy.

Plants generally don't care about where they get their nutrients, with the stipulation that these nutrients must be in appropriate soluble form and appropriate concentrations. Also, animals can eat any appropriate molecules regardless of source. Protein from a cow and the chemically identical protein from a bean are equally nutritious.

"Naturalness" also may carry the connotation of freedom from additives of all kinds. The use of additives in food is another place where ecological and medical considerations interact. Preservatives are useful precisely because they are poisonous to molds and bacteria. While we are biochemically different from molds and bacteria, all organisms are similar enough so that something that kills a bacterium is probably not the best thing for me, unless the bacterium itself is an even greater immediate danger.

Most foods may taste best when fresh, but some preservatives may enhance the flavor of foods while they ensure long shelf life. Salt meats and sausages, herrings, anchovies, sauerkrauts, and pickles owe much of their flavor to the salt that probably was first added as a preservative.

What are less defensible are some of the multisyllabic small-type dyes and other additives that are listed on the wrappings of ready-to-eat cereals, candy bars, breads, and other foods that are usually given to children. On the other hand, complete absence of these mystery additives may perhaps increase the average price of food and perhaps lower its quality.

It is possible to buy, at a premium price, vitamin C extracted from rose hips or from imported limes or any of a diversity of stylish "natural" sources. However, pure chemicals derived from organisms are not preferable to those created in other ways. Vitamin C synthesized by chemists is literally identical with that extracted from plants.

Similar arguments exist about "organic" fertilizers. Growing plants require specific chemicals, particular soil textures, appropriate amounts of water, and suitable light and temperature regimes. They generally don't care what kind of fertilizer supplies the chemicals they use.

Usually, but not always, soil bacteria and fungi help in producing soil textures, absorbing water and transmitting dissolved chemicals to plant roots. Nevertheless, as far as the chemistry itself is concerned, a molecule from a rotting leaf is identical to the same molecule made by a chemist.

Cautionary notes should be added. To say that chemically pure synthetic compounds are nutritionally identical with the same compounds derived from organisms seems true. The claim that complete diets of animals can be synthetically produced is generally false.

Our diets determine the kinds of bacteria to be found in our gut. There is a general belief that this explains the sometimes beneficial effects of consuming foods that are deliberately rich in bacteria, such as yogurt. It seems possible that heavy use of antibiotics in animal rearing and widespread use of plant products genetically engineered to contain antibiotics will have as yet unforeseen effects on our own bacterial populations.

Also, there are many trace components in our diets, and I believe that some of these are of value in ways that have not yet been identified. I take, and recommend to others, a position that avoids undue reliance on either chemists, mystics, or enthusiastic amateurs.

Agroindustrial farming relies heavily on fertilizers, pesticides, and fungicides. While these do not enter the food, so that the food is relatively safe, particularly if fruits and vegetables are washed and peeled. they do enter soils and watersheds and increasingly are producing major problems in drinking water supplies and fisheries. However, it is also possible to arrange the wastes from organic farming so that they are environmental hazards.

The use of different types of fertilizers is connected with particular political and economic systems. This is an issue of great importance, which I must omit from consideration.*

Only a few food plants and animals have not been genetically modified by selection. Brazil nuts, wild mushrooms, and ostriches come to mind, along with kangaroos and crocodiles raised for meat. Essentially all other farm-raised foods come from organisms that are genetically different from their wild relatives.

Often genetic changes are unintentionally selected in the process of domestication. Animals maintained by humans are continually being selected for tameness. Aggressive sheep and goats and cows are a nuisance and are eventually eliminated. Fast growers are usually selected for in agricultural conditions.

Should we be alarmed, from either a medical or ecological standpoint, about genetically "unnatural" plants and animals? Usually not, because there is hardly anything we eat whose genetic properties have not been strongly influenced by humans. However, there are some potential dangers.

One substitute for the use of chemical insecticides is the use of living bacteria capable of killing insects. *Bacillus thuringensis* is popular for this purpose. Recently seed companies have been offering farmers seeds in which genes from the bacteria have been incorporated into the corn or cotton genome, so that the

---

* It would require too much space and, because I am not expert in that field, too much time to gain sufficient expertise.

plants produce their own insecticides. These seeds are genetically engineered. They are not selected from the genetic variation present in the corn or cotton populations but are specifically manufactured to perform some particular job that is seen as commercially useful.

There is great concern that these genetically engineered seeds may have dangerous ecological properties that we cannot anticipate and that these consequences may set up problems more difficult and more expensive to solve than the initial problem of insect damage.

For example, what if a gene for immunity to insects is transmitted to pest plants? While we do not expect genes in corn to make their way into other species very often, examples of this type of genetic transmission are known. What if desirable species of insects are severely damaged by consuming engineered plants? How would this affect pollination of crops such as clover or of wildflowers?

The primary novelty of genetic engineering is that it can produce more rapid and more curious genetic modifications than could be produced by selecting genetic variants. Genes with specific properties can, in a sense, be injected directly into a genotype. I don't believe that the possible dangers of this process have yet been completely analyzed.

If you insist on the virtues of natural foods, you can seek out wild plants.[5] Many weeds are edible, but their use is limited. Berries and mushrooms can provide a breakfast picked while walking across campus. Public response to this sort of behavior is less than enthusiastic. However, chic restaurants do feature "weed salads." I am comfortable with wild salad, including delicious but not chic plants such as young chickweed.

Human genes can be placed into pigs, and presumably the converse will soon be possible. This has the possible advantage that the pig might be induced to make some enzyme or protein needed by certain humans. Conversely, a pig's gene introduced into a human cell might produce some required chemical for a person who is missing that genetic capacity.

I can imagine in a not-too-distant future the treatment of diabetes by inserting genetic insulin production systems into a human, thereby eliminating the need for injections. As far as I know, it is not now possible, but I would not be absolutely startled if it were possible by the time this book sees print.

We can in principle get results that might be dangerous. For example, a peanut gene in a corn plant might cause illness in those allergic to peanuts, or a gene for resistance to weed killers in a crop plant might under very special circumstances be lodged in a weed, which would then be immune to chemical control.

There are social, ecological, and political problems, but on the whole the health of human populations has been enhanced by the development of agricultural chemistry and genetics. Training or practice as an ecologist does not automatically provide the wisdom needed in considering these problems.

MEDICINE AND ECOLOGY AS "HEALTH" SCIENCES

Ecologists have nothing corresponding to the National Institutes of Health to support their research, so that one function of the leaders of ecological institutions is raising money. Finding relations between ecology and medicine might help.

A recent fund-raising letter from the director of an ecology institute contained the valid but self-serving information that "human infectious diseases are inherently ecological processes, because they involve complex webs of interactions among many species. An understanding of these ecological interactions can result in predictive power concerning disease risk. The science of ecology is an ally of health sciences, facilitating avoidance and prevention of disease."[6] He is correct, although the study of disease organisms is not the sole focus of ecology. There are also other similarities between ecology and medicine.

Ecology and medicine are both concerned with problems of survival of living systems. Both deal with issues of life, death, and health on the basis of incomplete information, limited theory, and imperfect techniques.

The idea of health in medicine refers to the well-being of single patients or, in the sense of public health, some average state of physical well-being of the individuals in a population. Health in an ecological sense also refers to individual organisms or single populations but is additionally used in the sense of the health of some assemblage of species or of a region or landscape. It is sometimes asserted, for example, that a region with a greater number of species in it is ecologically healthier than one with a smaller number of species. Although many ecologists may feel a personal desire to keep ecological systems healthy, the word *health* in this context may have a special meaning within the context of current ecology.

There was a great deal of good biology in ecology's long prehistory, when problems were recognized and often solved without any reference to a named profession or a scientific vocabulary. Crop rotation and fertilization, fallow field systems, careful placement of fields and houses, avoidance of stream contamination, and a wonderful capacity to domesticate a diversity of animals and plants all are examples of folk ecology, along with the expertise of fishing, hunting, and gathering of wild plant foods and fungi.

There is also medical prehistory. As long as three thousand years ago hospitals and temples were endowed to facilitate application of medical knowledge, to isolate certain kinds of illnesses, and to gain knowledge that might eventually be clinically useful.

In a modern context, physiology, biochemistry, cell biology, and the rest of the fascinating research areas that medical schools refer to as "basic sciences" are not immediately concerned with patient care but are providing the information from which, it may be hoped, procedures for patient care will be developed.

Some of the best of the early aggressive medical techniques were learned on battlefields. Manipulative operations with fair chances of patient survival consisted of amputations, splinting of broken arms and legs, and cauterizing of dirty wounds. In addition, interacting with academic medicine were the traditions of veterinary medicine and midwifery.

When writing became prevalent, medical literature rivaled in volume the tales of gods and heroes. Some medications and medical procedures are ancient. Among effective ancient medicines are alcohol, tea, coffee, and opium. Rhubarb and other herbs were used by folk practitioners as well as by academically trained physicians. Willow bark contains a chemical very similar to the active agent in aspirin and relieves fevers and headaches. I was recently prescribed a synthetic drug copied from a derivative of the foxglove plant. Dioscurides, physician in Nero's army, had noted foxglove's effect on the heart (in fact, it had been expected to have an effect on the heart because the flowers were "signed" with droplets that looked like blood*), and it has been part of traditional medicine for millennia.[7]

For the last several thousand years medicine has provided generally good advice on some ecological aspects of public health. There are restrictions on waste disposal in the Code of Hammurabi and the Bible.

---

* That statement is nonsense but the medicine works. What, if anything, can be inferred from that?

The ancients suggested that people avoid certain unhealthy areas. These were known as malarial places, places with "bad air" that produce fevers.

There has long been a general knowledge of dietetics. A twelfth-century Spanish physician states that meals should begin with salads and that if appetite can be assuaged by salad, that is best; meat may, however, be eaten after the salad course if one is still hungry.[8]

New medicines were imported to Europe in the sixteenth century by the same explorers who were bringing back ecological information. Quinine, from the bark of a South American tree (Jesuit bark), really is effective for malaria. Most exotic medicines were less useful than quinine. Nevertheless, there was a common feeling that things that were sufficiently strange, or even sufficiently disgusting, just might have some medicinal virtues. Even powdered mummies from Egyptian tombs were considered medicinal.[9] A visit to any modern herbal medicine shop, natural food store, or alternative medicine clinic indicates that this feeling persists.

Harmless folk remedies of dubious value are still in occasional use. These were generally based on plausible-sounding but essentially nonsensical quasi-theories—strings of verbal argument that mimicked normal science.

For example, when I was less than three years old I was taken to visit my great-grandmother on Manhattan's Lower East Side. The only memorable feature of the visit was the sight of a live piglet in a cardboard box under the stove. The purpose of the piglet (in a kosher kitchen!) was to attract rheumatism into itself and draw it from my great-grandmother.

In 1954 my American-born mother opened every drawer, door, and window in the house when my wife was in labor with our twins, to encourage "things to be open." If I had asked her point blank if she thought this would actually help, I am reasonably sure she would not have claimed that it worked. However, what harm could it do?

Four hundred years earlier, the physician Paracelsus advocated placing a sharpened axe under the bed during labor to cut pain, thereby replacing a series of unguents, bandages, and pills that might have been really harmful.[10] Bleeding, heavy doses of dangerous poisons (such as antimony, mercury, and arsenic), and bits and pieces of animals designed to cure by sympathetic magic (including the pulverized genitals of bears and tigers and powders made of rhinoceros horns) were certainly useless, very expensive, and sometimes harmful or even lethal.

There were strong disputes among various schools of medicine. Physicians who took very small amounts of blood when bleeding a patient were opposed by those who took appreciable amounts of blood. In this case, there were two variants of a useless procedure, each with its strong advocates. The argument mattered because one procedure was more harmful than the other. Similar arguments existed about dosage of medicinal poisons based on what were taken to be theoretical differences, with homeopaths advocating tiny doses and extreme dilution and mainstream medicine offering more dangerous dosages. The twentieth-century quarrels among psychoanalytic schools, before the development of neurological medications, had the same flavor.

We still see metaphysical arguments in medicine in the various approaches to alternative medicine. The term usually applies to medical practices that have not been part of the Western medical curriculum and have not been subject to statistically significant analyses. Mainstream medicine tends to ignore them or denigrate them. Their advocates attribute this to prejudice and ignorance.

We can find, in ecology, a list of horrors that may not rival those of medicine but stand up reasonably well on their own.

Most of these are recent, and many of them relate to attempts to imagine metaphysical organizing forces in nature, either in the teeth of the facts or merely as toothless philosophizing.[11] We have dealt with some of these in earlier chapters.

Very little helpful aggressive medicine was possible before the nineteenth century. Only the development of anesthesia and antiseptics permitted most kinds of surgery. Nevertheless, for millennia physicians, like clerics, were supported largely on faith.

Ecological decisions must often be made without the benefit of full-scale controlled experimentation. We do not have enough oceans for statistical testing, and large-scale treatments are too expensive to replicate.

Most medical practice occurs on the scale of a single patient. Treatment is a matter of an expert, the physician, interacting with a problem, the ill patient. Public health medicine and epidemiology take populations, rather than individuals, as the unit for treatment—in a sense the population is the patient. There usually is no public debate over the treatment of a single patient, except if the patient is politically important. Public health problems, on the other hand, generate public dispute. Ecology concentrates on populations of organisms rather than single organisms. Disputes are very common in applying ecology.

Problems of water quality are prime examples of interaction between medical and ecological issues. A hundred years ago Ibsen's play *The Enemy of the People* described the political and social repercussions of a medical decision about water purity, specifically the closing of a swimming beach in the summer. Since then the fields of environmental law and public health have arisen in response to this class of disputes. Both of these may deal with individuals, but their interesting effect is on a population of individuals rather than the single individual.

One great difference between medicine and most of ecology

is how they use the idea of death. Most physicians see the death of their patient as the enemy. Ecologists distinguish between different kinds of death. Optimal medical or ecological solutions for a population frequently conflict with the needs of individuals in that population. This is seen in the use of triage.

In its strong form, triage occurs only in military contexts or disaster situations, when the medical needs exceed the immediately available medical resources. Patients are divided into three categories. Some have relatively minor problems, or at least problems whose treatment can be postponed for a while. Others must be treated as soon as possible to avoid profound deterioration, and others have been so badly damaged that treatment would be of no avail. This third category is "black-tagged" for comfort care only, in anticipation of death.

By efficient triage, medical resources are not wasted on the dying or those in no immediate danger. This permits concentration of activity where it can do the most good.

A patient entering an emergency room will initially be seen by a triage nurse, but in normal times there is no threat of a black tag:

> The concept of triage in everyday emergency medicine is simply the determination of how long a patient can wait. . . . What actually happens to conserve resources is that certain ailments are "black-tagged" as not being appropriate for the [emergency room] to address. Chronic psychiatric problems, drug addiction, homelessness, alcoholism, dental problems, HIV infection, and tobacco dependence would all fit into that category. These problems [are dealt with] by writing a referral, and the patient is politely discharged after his "acute," i.e. "acceptable," problems have been dealt with. This frequently happens over and over again with the same patient.[12]

Ecological problems are also seen as subject to a type of triage. Attending to small-scale problems with a possibility for rapid deterioration, such as changes in population sizes, call for immediate response, while global warming used to be handled as if treatment could wait, and tectonic changes are beyond us.

Perhaps one distinction between early medical errors and those that were committed by the precursors of ecology is that medical errors were often due to profound errors in the theoretical understanding of physiology and anatomy, while ecological errors arose from a general desire to do good, without any deep theory at all.

Although physicians focus on one species, many other species are also studied in the context of human medicine. Bacteria and yeasts can be good model organisms for human biochemistry. Some mammals develop ailments similar to those of people and are therefore studied in detail. Popular experimental animals include rats, mice, and guinea pigs, but rabbits, apes and monkeys, and even armadillos (in the case of leprosy) have been used as models for diseased humans. Veterinarians study perhaps twenty more mammals in addition to other organisms that are useful, such as birds and fish. Generally, the richness of knowledge about humans is greater than that about other organisms.

Ecologists are concerned with hundreds of species, from protozoa to flying foxes, and also with the air and water and soil that surround them. More than a million species are known, but the vast preponderance are not the focus of any immediate attention. In most cases the knowledge of any one of these organisms is considerably less than that possessed by veterinarians or physicians about their test animals.

Applied ecologists find practical problems on many levels. If there is concern for a particularly rare or endangered species, it may be necessary to rear individual organisms. Ecology then resembles a kind of veterinary medicine.

Until recently, the basic-science aspects of ecology have been the most important focus of effort by academic ecologists. These basic studies include natural history of plants and animals, animal behavior, biogeochemistry, marine biology, evolution, and more. They are fascinating and appeal strongly to young students. The number of students who profess a vocation for marine biology is startling.

Many ecologists now believe that certain species, certain landscapes, and certain rivers and oceans cry out for treatment of some sort. Ecologists claim that they are the ones to make the triage decisions and to provide the treatments. The truth of this statement is a center of concern. There is an enormous range of opinions about what is or is not to be done in any particular case. Not all opinions can be correct.

Solving ecological problems involves governmental activity, major financial commitment, and sometimes even international accords, or, failing that, wars. While it is not possible to deal with all ecological problems, the world will be tragically different if no attempt is made to deal with any of them. For the moment the decisions are confrontational and the ecologists act as experts.

Is the triage problem inevitable? Isn't it possible that some technological breakthroughs will alter, say, the water shortage problem, and all concerned could be satisfied? What if cheap desalination were available? In general, helpful technology must either lower the demand for a resource or increase the quantity of the resource if the general problem is to be alleviated. Wouldn't unregulated demand rise to meet the new supply, and wouldn't regulation of demand require allotment decisions?

Ultimately ecological information alone does not permit

decision. Consideration must be given to social, financial, legal, and other issues. However, absence of ecological information or misinterpretation of ecological information may also make environmental decisions almost impossible.

Even in the situation of abundant empirical information, social decisions require ethical considerations, if not necessarily ethical conclusions.[13]

Ethical problems must involve triage. The relative value of competing needs must be decided. Definition and study of ethical issues may occur in an academic setting, but the only way to advance toward practical decisions about ethical questions is through public discussion. That is why applied ecology cannot avoid being political.

The question becomes how ecological assertions can be both politically and scientifically acceptable. The answers vary with the details of each problem. There is no general answer.

Conclusions

I hope this book will be a step toward what Bruce Wallace has called "ecological literacy."[1] Literacy does not tell you what you ought to read or what you ought to believe about any particular issue, nor does education consist of memorizing a list of problems and their solutions. We cause changes, but we are not condemned to life in endless urban slums unless we behave like boorish illiterates.

A literate person confronted with any problem can analyze the situation and then try to find a best next step. The choice of goals dictates the problems that must be solved to reach those goals. If the problems are insurmountable, perhaps the goals will have to be changed.

Organisms, throughout their billion-year history, have had to respond to a broad spectrum of disturbances. Can we learn from them?

Elton made the powerful observation many years ago that when animals are carefully observed, it turns out that most often they are doing nothing, or at any rate "nothing in particular."[2]

Once I watched a lion at a water hole sleeping near a herd

of zebras. Suddenly the lion woke up and made a rush at the zebras. The zebras immediately trotted off a few yards. If the lion had continued to rush at them, they would have galloped off at full tilt. In fact, the lion returned to sleeping and the zebras to grazing and drinking. The presence of the zebras stimulated the lion to a weak attack. The rush by the lion turned out to be a small disturbance for the zebras. Had the lion been extremely hungry, the chase and the flight would have been more determined.

Determined hunting by a pride of lions results in several minutes of running by both predator and prey, often terminating when the lions tire. Sometimes a lion actually springs at a zebra. The attacked zebra usually succeeds in running away. We can imagine that this particular zebra will be more wary in the future. The zebras that do not escape are what keep the food chain going.

Any organism has a series of responses to changes. These responses differ in timing and in how much of a commitment the organism makes to that response. Generally, what may be a small-scale, temporary problem is best responded to by a small-scale, reversible response mechanism while having in reserve the possibility that what is thought of as a small-scale disturbance might persist and grow larger.

A similar pattern of responses to a new disturbance can be found in almost any organism. The immediate response is as if to a minor and temporary problem. The later responses involve a deeper commitment of resources by the organism and come into play only if the problem has not gone away. The process of evolution modifies the response processes in accord with the recent history of disturbances.[3]

Even among plants, the response to, say, a lack of water involves a series of steps that can still be ordered in terms of their speed of recovery and depth of commitment. A hot summer day may cause leaves to wilt, but they may recover after the sun has set.

Gradation of responses would seem ideal for living in an uncertain world in which one is endowed with only limited resources and a potentially infinite number of possible problems. Unfortunately, there is no general evolutionary mechanism for adjusting the pattern of responses by a human society to environmental disturbances. As occurs so often, we are thrown back on our own political and intellectual powers. Nevertheless, we can derive some hints from the response patterns of organisms.

Small quasi-ecological problems abound, such as raccoons in the garage, basement floods, broken plumbing, roof leaks, trash, worn-out rugs, and rotting window frames. My wife and I have dealt with all of these this summer. In these examples the deterioration is of private property and the advantages of good responses immediately accrue to my family.

Problems of zoning, garbage collection, park maintenance, and insect spraying are usually dealt with by agencies of local government. Usually the people involved agree with what is being done. Sometimes they do not. Disputes between individuals are settled by argument, complaints to police, and perhaps lawsuits.

Perhaps the most serious ecological changes are those occurring on large geographical and temporal scales. Water and air pollution, global temperature change, land use change, and species extinction are the most prominent. These need not make any obvious and immediate difference to any particular individual.

Sometimes the responses of organisms to dangers may themselves become dangerous. It has been suggested that fevers in response to bacterial infections are a way of destroying the bacteria.[4] Sometimes fevers that occur in response to infections are so severe as to endanger other life functions. Similarly, the attempts to solve large ecological problems may seem to be more of a nuisance than the problems themselves.

Ultimately solutions to the big problems will be necessary to avoid major human misery. The attempted solutions require expenditures of time, money, and resources and may involve curtailing freedom.

The political and economic heat surrounding these problems hinges on the possibility that some of the suggested solutions may be more damaging than the problems they are intended to cure or, conversely, that failure to actively address these issues at the present time will exacerbate future difficulties. I cannot provide a general answer. As a society, we will have to find the answers and pay the cost of acting or not acting in the context of particular problems.

The only certain generalization is that ecology is not merely a fad or a subspecialty of biology. We are in it. It will not go away, however much fads may change.

Sometimes the parties in ecological disputes are entire nations, government agencies, or very large corporations, so that the possible role for individuals is severely limited. They are of the scale of serious politics. Disputes between large organized groups may be settled by public debates, elections, or even wars.

In short, ecological problems must be dealt with like other problems. First there must be clear factual information. Using that information may be difficult. Public discussions are time-consuming, and they will not always produce the results you want. Alternative systems may be worse. How questions are approached varies with the kind of government in power. Different systems respond to different approaches.

Many years ago I chaired an international symposium discussing how ecologists can set up protective reserves for endangered populations of organisms. Among the group were a few Englishmen who proudly described how they enlisted the interest of the local landowners, who posted land to prevent hunting. There were some Americans describing how they convinced local legislatures to enact special legal rulings to preserve particularly vulnerable areas. An Israeli talked about how an educational program had been established in the local schools so that children would not damage wildflowers. A Venezuelan then said that this all seemed time-consuming and uncertain. What he had done, as part of a project to protect capybaras from poachers, was to contact the local police chief and explain the problem. Immediately armed guards appeared and the poaching danger vanished.

How we argue about ecology is often as important as the substance of our arguments. There is a temptation to focus on winning the argument. The issues of ecology are too important for that. If you happen to win an important argument by using what you know to be bad science, in the long run you will have damaged the future of science itself. You will have destroyed the authority of good science, and on some future occasion opponents may use equally specious science to destroy your position. There is a danger that using bad science to effectively win arguments will cost us one of our only sources of clear truths.

Disputes are inevitable and do not necessarily imply that your opponents are evil, although it doesn't preclude that possibility. Essentially all decisions will be compromises. All ecological decision making is contingent on mutual understandings, political power, economic needs, and desires as well as on empirical circumstances in nature. Even the best of scientific arguments are often not sufficient for solving particular problems.

A participant in a public debate can, and perhaps must, use law and lawyers as weapons. The consequences of this are not simple.

Questions involving environmental decision or legal action usually require case-by-case detailed analysis. We must continue to muddle through on most ecological issues as we do on preserving endangered species—lawsuit by lawsuit, town meeting by town meeting, election by election, habitat by habitat.[5] The job appears endless, but it cannot be postponed or ignored. You, or delegates you trust, must protect your habitat, or it will not meet your needs and desires.

If issues are not made public, you may not receive fair treatment. For example, in the early seventies I was on an advisory committee for ecological problems in New York City. A high-level city official explained how the flight paths of aircraft were chosen. As far as possible, flights were routed over Harlem for the plausible reason that the people of Harlem did not generally complain about airplane noise as loudly as people from other neighborhoods did.

He was not being evil or bigoted. He was merely doing his day's work in the most convenient way that he could. I don't actually know if that sort of thing continues, but whoever you are and wherever you live, it is worth an attempt to find out how decisions of this sort are made.

Political, technological, economic, and sometimes moral change cannot be automatically reversed or obliterated in the name of ecology. As the spatial and temporal scale of a problem becomes larger it becomes more and more difficult to be fair to all the people and other organisms involved. Who is benefiting and by how much, who pays or loses and how much, and how do we know? What are the broader ecological consequences of decisions?

Some people enjoy being activists. Most do not. Nevertheless, you are a participant in ecologically significant disputes whether you want to be or not. Perhaps ecological regulations will interfere with your livelihood or profit margin. Perhaps the desire for profit will change the standards for your air or water supply, lower the value of your house, corrupt the quality of your vacation, or sicken your children. You must decide how much these things concern you, and act accordingly.

Various movements advocate particular ways of looking at nature. It is possible to look at nature in ways that will make it very difficult to answer serious questions. For example, there is no deity called Nature that can label organisms or situations or changes as good or bad.

One certain conclusion is that we cannot assign moral categories to natural objects or events. We cannot assign goodness or badness to an organism. Scorpions and crocodiles and pit vipers are not lovable, nor are they vicious. Rabbits and sparrows and rosebushes are not benign, although they may be lovable. To designate some groups of species as "bad"—as is done, for example, in the debates on invasive species—is an invitation to nonsense.

To assign moral purpose to natural events is not helpful for management. The moral implications of management procedures are our problem, not nature's. We must guard and even construct the properties of our own world to meet our needs, desires, and moral concerns.[6] We are the only organisms capable of this in any serious way.

Will there be an intellectual breakthrough that will provide a general set of guidelines for all problems in ecology? Will the solution to ecological problems ever spring automatically from having a correct attitude? I don't think so.

Will there be a mass rejection of ecology that will wipe out ecological activism completely? This is not unrealistic. There are millions out there trying. It is an ongoing threat.

Will there be a green-oriented party that will actually run the government using admittedly inadequate science to legislate limits on use of property? Perhaps, but how will it be organized and justified? If it is not justified by facts or by finance, will it be justified by military force or metaphysics?

This book is incomplete. It could go on with more examples and problems, almost forever, or it can stop here. Certainly, the more we actually know about ecology and the more carefully we analyze the concepts underlying ecology, the better our chances of avoiding disasters and the greater our pleasure in our battered natural world.

Will we continue to muddle through? I hope so.

Appendix

The Rate of Increase of Populations

If we are interested in the value of R for some kind of organism in some set of circumstances, we might experimentally set up growing populations and wait to see what value of R is attained, but this is expensive in money and time.

It is possible to evaluate R from age-specific survival and mortality rates.

We define

L_x = the probability that an organism born x time units ago will be alive at the halfway mark between age x and $x + 1$

M_x = the number of young born to a mother of age x during the interval $x, x + 1$

R = the exponential increase rate of the population after it has achieved stable age distribution

L_x and M_x are directly observable if we know the age of each of the organisms and watch them over an appropriate time interval. If age is measured in days, observation of reproduction over one day of individuals x days old gives us M_x. We can write Equation 1, in which the only unknown term is R, as

Equation 1

$$1 = \sum L_x M_x R^{-x}$$

in which the summation is over the full range of possible ages.

It is also possible to evaluate the relative value to the population, in terms of how much contribution will be made to future growth, of organisms of different ages.

Individual reproductive rates and death rates as functions of age are combined in such a way that there is a single number, which we can approximate such that all the births and deaths combine to reproduce one animal—the one that was loaned the life in the first place. The equation tells us how much an organism is worth at different ages compared to a single newborn that is assumed to have a value of either 1 (if we consider only females) or 2 (if we consider parental pairs).

In addition to our previous definitions we define:

V_x = the reproductive value of an organism of age x

V_0 = the reproductive value of a newborn organism, which can be set at 1 or 2.

The equation is then:

$$V_y / V_o = (R^y (\Sigma_{y \text{ to infinity}} L_y M_y))/L_y.$$

References

Introduction

1. L. Slobodkin, *Simplicity and Complexity in Games of the Intellect* (Cambridge: Harvard University Press, 1992).
2. P. S. Martin and R. G. Klein, eds., *Quaternary Extinctions: A Prehistoric Revolution* (Tucson: University of Arizona Press, 1984).
3. B. McKibben, *The End of Nature* (New York: Random House, 1989).
4. L. B. Slobodkin, "Ecosystems and the Maginot Line: Review of Frank Golley, 1993, A *History of the Ecosystem Concept in Ecology: More than the Sum of Its Parts*," TREE 9 (1994): 408–10; F. Golley, *A History of the Ecosystem Concept in Ecology: More than the Sum of Its Parts* (New Haven: Yale University Press, 1993); E. P. Odum, "Emergence of Ecology as a New Integrative Discipline," *Science* 195 (1977): 1289–93.
5. D. Worster, *Nature's Economy: A History of Ecological Ideas* (Cambridge: Cambridge University Press, 1985); R. P. McIntosh, *The Background of Ecology: Concept and Theory* (New York: Cambridge University Press, 1985).
6. T. R. Malthus, *An Essay on the Principle of Population As It Affects the Future Improvement of Society, with Remarks on the Speculations of Mr. Godwin, M. Condorcet, and Other Writers* (London: J. Johnson, 1798); T. R. Malthus, *Observations on the Effects of the Corn Laws, and of a Rise or Fall in the Price of Corn on the Agriculture and General Wealth of the Country* (London: J. Johnson and Co., 1814).

7. P. Ehrlich, *The Population Bomb* (New York: Ballantine, 1968).
8. J. E. Cohen, *How Many People Can the Earth Support?* (New York: W. W. Norton, 1995).
9. P. Sears, *Deserts on the March* (Norman: University of Oklahoma Press, 1935).
10. J. Sheail, *Seventy-five Years in Ecology* (London: British Ecological Society, 1988); L. B. Slobodkin, "75 Years in Ecology—Review of Sheail, J. *Seventy-five Years in Ecology*," *Science* 242 (1988): 784–85.
11. R. Carson, *Silent Spring* (Boston: Houghton Mifflin, 1962).
12. E. Abbey, *The Monkey Wrench Gang* (New York: Avon Books, 1992).
13. D. Barry and A. Baker, "'Eco-Terrorism' Group, a Hidden Structure and Public Message," *New York Times*, January 14, 2001, B1, B5.
14. Ehrlich, *The Population Bomb.*
15. R. Gelbspan, *The Heat Is On: The High Stakes Battle Over Earth's Threatened Climate* (New York: Addison-Wesley, 1997).
16. Malthus, *Observations on the Effects of the Corn Laws.*
17. B. Wallace, *The Environment: As I See It, Science Is Not Enough* (Elkhorn, W. Va.: Elkhorn Press, 1998).
18. B. Russell, *Unpopular Essays* (New York: Simon and Schuster, 1950).
19. P. Errington, *Muskrat Populations* (Ames: Iowa State University Press, 1963); P. Errington, *Of Men and Marshes* (New York: Macmillan, 1957).
20. H. D. Thoreau, *Faith in a Seed* (Washington, D.C.: Island Press, 1993).
21. Wallace, *The Environment.*
22. R. MacArthur and J. MacArthur, "On Bird Species Diversity," *Ecology* 42 (1961): 594–98.
23. L. Slobodkin, F. Smith, and N. Hairston, "Regulation in Terrestrial Ecosystems and the Implied Balance of Nature," *American Naturalist* 101 (1967): 109–24; N. Hairston, F. Smith, and L. Slobodkin, "Community Structure, Population Control and Competition," *American Naturalist* 94 (1960): 421–25.

Chapter 1: The Big Picture

1. W. Whewell, *Astronomy and General Physics Considered with Reference to Natural Theology. Treatise Three* (Philadelphia: Carey, Lea,

and Blanchard, 1833); L. J. Henderson, *The Fitness of the Environment: An Inquiry into the Biological Significance of the Properties of Matter* (New York: Macmillan, 1913).

2. R. A. Griffiths, "Temporary Ponds as Amphibian Habitats," *Aquatic Conservation—Marine and Freshwater Ecosystems* 7 (1997): 119–26.
3. A. L. Lavoisier, *La chaleur et la respiration, 1770–1789* (Paris: Masson, 1892).
4. L. Slobodkin and S. Richman, "The Calories/Gram in Species of Animals," *Nature* 191 (1961): 209.
5. L. Slobodkin, "Energy in Animal Ecology," *Advances in Ecology* 1 (1962): 69–101.
6. L. E. Orgel, "The Origin of Life—A Review of Facts and Speculations," *Trends in Biochemical Sciences* 23 (1998): 491–95.
7. H. J. Morowitz et al., "The Origin of Intermediary Metabolism," *Proceedings of the National Academy of Sciences of the United States of America* 97 (2000): 7704–8.
8. B. Rasmussen, "Filamentous Microfossils in a 3,235-Million-Year-Old Volcanogenic Massive Sulphide Deposit," *Nature* 405 (2000): 676–79.
9. L. Slobodkin et al., "A Review of Some Physiological and Evolutionary Aspects of Body Size and Bud Size of Hydra," *Hydrobiologia* 216/217 (1991): 377–82; P. Bossert and K. Dunn, "Regulation of Intracellular Algae by Various Strains of the Symbiotic *Hydra viridissima*," *Journal of Cell Science* 85 (1996): 187–95; P. Bossert, "The Effect of Hydra Strain Size on Growth of Endosymbiotic Alga," Ph.D. thesis, State University of New York at Stony Brook, 1988; L. Slobodkin and P. Bossert, "Cnidaria—or Coelenterates," in D. Thorp and D. Covitch, eds., *The Freshwater Invertebrates* (New York: Academic Press, 2001), 135–54.
10. Y. Loya and R. Klein, *Shonit ha'Almogim* (Tel Aviv: Israeli Ministry of Defense, 1994).
11. H. D. Holland and N. J. Beukes, "A Paleoweathering Profile from Griqualand West, South-Africa—Evidence for a Dramatic Rise in Atmospheric Oxygen Between 2.2 and 1.9 Bybp," *American Journal of Science* 290A (1990): 1–34.
12. T. M. Lenton and A. J. Watson, "Redfield Revisited 2. What Reg-

ulates the Oxygen Content of the Atmosphere?" *Global Biogeochemical Cycles* 14 (2000): 249–68.

13. J. M. Moldowan, J. Dahl, et al., "The Molecular Fossil Record of Oleanane and Its Relation to Angiosperms," *Science* 265, 5173 (1994): 768–71.
14. G. Hutchinson, *A Treatise on Limnology*, Volume 1: *Geography, Physics and Chemistry* (New York: John Wiley and Sons, 1957).
15. A. S. Cohen, M. J. Soreghan and C. A. Schotz, "Estimating the Age of Formation of Lakes—An Example from Lake Tanganyika, East African Rift System," *Geology* 21, (1993): 511–14.
16. H. D. Thoreau, *Walden, or Life in the Woods* (Boston: Ticknor and Fields, 1854).
17. C. Darwin, *The Origin of Species by Means of Natural Selection* (6th edition London: Watts and Co., 1929).
18. B. Maguire, "The Passive Dispersal of Small Aquatic Organisms and Their Colonization of Isolated Bodies of Water," *Ecological Monographs* 33 (1963): 161–85.
19. N. Hairston Jr. et al., "Age and Survivorship of Diapausing Eggs in a Sediment Egg Bank," *Ecology* 76 (1996): 1706–11.
20. Griffiths, "Temporary Ponds as Amphibian Habitats."
21. G. Fryer and T. D. Iles, *The Cichlid Fishes of the Great Lakes of Africa: Their Biology and Evolution* (New York: Crown, 1972).
22. V. Takhteev, "Trends in the Evolution of Baikal Amphipods and Evolutionary Parallels with Some Marine Malacostracan Fauna," *Advances in Ecological Research* 31 (2000): 197–220.
23. G. Fryer, "Evolution and Adaptive Radiation in the Macrothricidae (Crustacea: Cladocera): A Study in Comparative Functional Morphology and Ecology," *Philosophical Transactions of the Royal Society of London, Series B—Biological Sciences* 269 (1974): 221–385.
24. C. Zimmer, *Parasite Rex: Inside the Bizarre World of Nature's Most Dangerous Creatures* (New York: Free Press, 2000).
25. C. R. Goldman, "Primary Productivity, Nutrients, and Transparency During the Early Onset of Eutrophication in Ultra-oligotrophic Lake Tahoe, California, Nevada," *Limnology and Oceanography* 33 (1988): 1321–33; C. R. Goldman, "Lake Tahoe—Preserving a Fragile Ecosystem," *Environment* 31 (1989): 6–14; T. F. Thingstad,

"A Theoretical Approach to Structuring Mechanisms in the Pelagic Food Web," *Hydrobiologia* 363 (1998): 59–72.

26. M. Goulding, *The Fishes and the Forest: Explorations in Amazonian Natural History* (Berkeley: University of California Press, 1980).
27. G. P. Glasby, "Earliest Life in the Archean: Rapid Dispersal of CO_2-Utilizing Bacteria from Submarine Hydrothermal Vents," *Episodes* 21 (1998): 252–56; T. B. Brill, "Geothermal Vents and Chemical Processing: The Infrared Spectroscopy of Hydrothermal Reactions," *Journal of Physical Chemistry A* 104 (2000): 4343–51.
28. L. Slobodkin, "A Possible Initial Condition for Red Tides on the Coast of Florida," *Sears Journal of Marine Research* 12 (1953): 148–55.
29. S. Ferson, "Are Vegetation Communities Stable Assemblages?" Ph.D. thesis, State University of New York at Stony Brook, 1988.
30. N. Hairston and G. Byers, "The Soil Arthropods of a Field in Southern Michigan: A Study in Community Ecology," *Contributions from the Laboratory of Vertebrate Biology, University of Michigan* 64 (1954): 1–37.
31. D. E. Dykhuizen, "Santa Rosalia Revisited: Why Are There So Many Species of Bacteria?" *Antonie Van Leeuwenhoek International Journal of General and Molecular Microbiology* 73 (1998): 25–33.
32. R. M. Auge, "Water Relations, Drought and Vesicular-Arbuscular Mycorrhizal Symbiosis," *Mycorrhiza* 11 (2001): 3–42; L. Brussaard, T. W. Kuyper, and R.G.M. de Goede, "On the Relationships Between Nematodes, Mycorrhizal Fungi and Plants: Functional Composition of Species and Plant Performance," *Plant and Soil* 232 (2001): 155–65; H. Roussel et al., "Signaling Between Arbuscular Mycorrhizal Fungi and Plants: Identification of a Gene Expressed During Early Interactions by Differential RNA Display Analysis," *Plant and Soil* 232 (2001): 13–19; F. Martin, "Frontiers in Molecular Mycorrhizal Research—Genes, Loci, Dots and Spins," *New Phytologist* 150 (2001): 499–505.
33. Zimmer, *Parasite Rex.*
34. W. G. Whitford, "The Importance of the Biodiversity of Soil Biota in Arid Ecosystems," *Biodiversity and Conservation* 5 (1996): 185–95.
35. D. Thompson, *On Growth and Form* (Mineola, N.Y.: Dover Press, 1992).

36. S. D. Johnson, "Batesian Mimicry in the Non-Rewarding Orchid *Disa pulchra*, and Its Consequences for Pollinator Behaviour," *Biological Journal of the Linnean Society* 71 (2000): 119–32; F. P. Schiestl et al., "Sex Pheromone Mimicry in the Early Spider Orchid (*Ophrys sphegodes*): Patterns of Hydrocarbons as the Key Mechanism for Pollination by Sexual Deception," *Journal of Comparative Physiology A—Sensory, Neural and Behavioral Physiology* 186 (2000): 567–74.
37. M. F. McKenna and G. Houle, "Why Are Annual Plants Rarely Spring Ephemerals?" *New Phytologist* 148 (2000): 295–302.
38. Thoreau, *Walden;* S. Forbes, "The Lake as a Microcosm," *Bulletin,* Peoria (Ill.) Scientific Association, 1887, 77–87 (reprinted in *Bulletin,* Illinois Natural History Survey 15 [1925]: 537–50).
39. N. Wiener, *Cybernetics, or Control and Communication in the Animal and the Machine* (New York: J. Wiley, 1948); G. Hutchinson, "Circular Causal Systems in Ecology," *Annals of the New York Academy of Science* 50 (1948): 221–46.
40. R. A. Armstrong et al., "A New, Mechanistic Model of Organic Carbon Fluxes in the Ocean Based on the Quantitative Association of POC with Ballast Minerals," *Deep Sea Research II* 7 (2001): 1–18; R. A. Armstrong and R. A. Jahnke, "Decoupling Surface Production from Deep Remineralization and Benthic Deposition: The Role of Mineral Ballasts," *U.S. JGOFS News* 11 (2001): 1–2.

Chapter 2: How Do Species Survive?

1. K. Steudel, W. P. Porter, and D. Sher, "The Biophysics of Bergmann's Rule—A Comparison of the Effects of Pelage and Body-Size Variation on Metabolic-Rate," *Canadian Journal of Zoology—Revue Canadienne de Zoölogie* 72 (1994): 70–77; C. B. Marcondes et al., "Influence of Altitude, Latitude and Season of Collection (Bergmann's Rule) on the Dimensions of *Lutzomyia intermedia* (Lutz & Neiva, 1912) (Diptera, Psychodidae, Phlebotominae)," *Memorias do Instituto Oswaldo Cruz* 94 (1999): 693–700; J. R. Bindon and P. T. Baker, "Bergmann's Rule and the Thrifty Genotype," *American Journal of Physical Anthropology* 104 (1997): 201–10; K. G. Ashton, M. C. Tracy, and A. de Queiroz, "Is

Bergmann's Rule Valid for Mammals?" *American Naturalist* 156 (2000): 390–415.

2. C. R. Marshall, E. C. Raff, and R. A. Raff, "Dollo's Law and the Death and Resurrection of Genes," *Proceedings of the National Academy of Sciences of the United States of America* 91 (1994): 12283–7; M.S.Y. Lee and R. Shine, "Reptilian Viviparity and Dollo's Law," *Evolution* 52 (1998): 1441–50; J.C.V. Klein, "Interpretation of Character Phylogenies in Calanoid Copepods by Implementing Dollo's Law," *Journal of Crustacean Biology* 18 (1998): 153–60; S. J. Gould and B. A. Robinson, "The Promotion and Prevention of Recoiling in a Maximally Snaillike Vermetide Gastropod—A Case-Study for the Centenary of Dollo's Law," *Paleobiology* 20 (1994): 368–90.
3. C. Elton, *Animal Ecology* (Oxford, 1927), 64.
4. F. Nottebohm et al., "The Life-Span of New Neurons in a Song Control Nucleus of the Adult Canary Brain Depends on Time of Year When These Cells Are Born," *Proceedings of the National Academy of Sciences of the United States of America* 91 (1994): 7849–53.
5. A. Huxley, *After Many a Summer Dies the Swan* (New York: Harper and Brothers, 1939); R. Firbank, *Valmouth: A Romantic Novel* (London: Grant Richards Ltd., 1919); A. Tennyson, "Tithonus," in I. Lancashire, ed., *Representative Poetry On-line, 1997* (Toronto: Web Development Group, University of Toronto Library, 1997), 1860.
6. L. Slobodkin et al., "A Review of Some Physiological and Evolutionary Aspects of Body Size and Bud Size of Hydra," *Hydrobiologia* 216/217 (1991): 377–82.
7. D. Martinez, "Mortality Patterns Suggest Lack of Senescence in Hydra," *Experimental Gerontology* 33 (1998): 217–25.
8. L. Slobodkin, *Growth and Regulation of Animal Populations* (New York: Dover Publications, 1980).
9. D. Lack, *The Natural Regulation of Animal Numbers* (Oxford: Clarendon Press, 1954).
10. W. J. Loughry et al., "Polyembryony in Armadillos," *American Scientist* 86 (1998): 274–79.
11. J. E. Cohen, *How Many People Can the Earth Support?* (New York: W. W. Norton, 1995).

12. C. Simon et al., "Genetic Evidence for Assortative Mating Between 13-year Cicadas and Sympatric '17-Year Cicadas with 13-Year Life Cycles' Provide Support for Allochronic Speciation," *Evolution* 54 (2000): 1326–36; K. Heliovaara, R. Vaisanen, and C. Simon, "The Evolutionary Ecology of Insect Periodicity," *Trends in Ecology and Evolution* 9 (1994): 475–80.
13. L. Slobodkin, "Population Dynamics in *Daphnia obtusa Kurz*," *Ecological Monographs* 24 (1954): 69–88.
14. N. Hairston Jr. et al., "Age and Survivorship of Diapausing Eggs in a Sediment Egg Bank," *Ecology* 76 (1996): 1706–11.
15. R. F. Heizer and T. Kroeber, eds., *Ishi the Last Yahi: A Documentary History* (Berkeley: University of California Press, 1979).
16. S. M. Shultz et al., *Conservation Biology: With RAMAS EcoLab* (Sunderland, Mass.: Sinauer Associates, 1999).
17. Slobodkin, *Growth and Regulation of Animal Populations*.
18. A. Murie, *The Wolves of Mt. McKinley* (Washington, D.C.: U.S. Department of the Interior, 1944).
19. A. Lotka, *Elements of Physical Biology* (New York: Dover Press, 1956).
20. R. A. Fisher, *The Genetical Theory of Natural Selection* (New York: Dover Press, 1958).
21. M. Burgman and D. Lindenmayer, *Conservation Biology for the Australian Environment* (Chipping Norton, NSW: Surrey Beatty and Sons, 1998).
22. F. E. Smith, "Quantitative Aspects of Population Growth," in E. Boell, ed., *Dynamics of Growth Processes* (Princeton: Princeton University Press, 1954).
23. I. N. Wang, D. E. Dykhuizen, and L. B. Slobodkin, "The Evolution of Phage Lysis Timing," *Evolutionary Ecology* 10 (1996): 545–58.
24. L. Slobodkin, "Energy in Animal Ecology," *Advances in Ecology* 1 (1962): 69–101.
25. L. Slobodkin, "How to Be a Predator," *American Zoologist* 8 (1968): 43–51; L. Slobodkin, "Prudent Predation Does Not Require Group Selection," *American Naturalist* 108 (1974): 665–78.
26. Y. Loya and L. Slobodkin, "The Coral Reefs of Eilat (Gulf of Eilat, Red Sea)," in D. R. Stoddard and M. Yonge, eds., *Regional Variation*

in Indian Ocean Coral Reefs (Bristol: Academic Press, 1971); L. Slobodkin and L. Fishelson, "The Effect of the Cleaner-Fish *Labroides dimidiatus* on the Point Diversity of Fishes on the Reef Front at Eilat," *American Naturalist* 108 (1974): 369–76; Y. Loya and R. Klein, *Shonit ha'Almogim* (Tel Aviv: Israeli Ministry of Defense, 1994).

27. D. B. Botkin et al., "A Foundation for Ecological Theory," *Memoria dell'Istituto Italiano di Idrobiologia,* 37 supp. (1979): 13–31.
28. Slobodkin and Fishelson, "The Effect of the Cleaner-Fish *Labroides dimidiatus.*"
29. A. Beattie and P. Ehrlich, *Wildsolutions: How Biodiversity Is Money in the Bank* (New Haven: Yale University Press, 2001).
30. N. G. Yoccoz, J. D. Nichols, and T. Boulinier, "Monitoring of Biological Diversity in Space and Time," *Trends in Ecology & Evolution* 16 (2001): 446–53.
31. W. Clinton, executive order on invasive species, February 3, 1999.
32. S. Levin, ed., *Encyclopedia of Biodiversity* (New York: Academic Press, 2001).
33. D. C. Adams et al., "An 'Audience Effect' for Ecological Terminology: Use and Misuse of Jargon," *Oikos* 80 (1997): 632–36.
34. Office of Technology Assessment, U.S. Congress, *Preparing for an Uncertain Climate* (Washington, D.C.: Government Printing Office, 1993); Office of Technology Assessment, U.S. Congress, *Changing by Degrees: Steps to Reduce Greenhouse Gases* (Washington, D.C.: Government Printing Office, 1991); Office of Technology Assessment, U.S. Congress, *Technologies to Maintain Species Diversity* (Washington, D.C.: Government Printing Office, 1987).
35. A. J. Lotka, *Elements of Physical Biology* (Baltimore: Williams and Wilkins, 1925); A. J. Lotka, "Théorie analytique des associations biologiques," *Actualités Scientifique et Industrielles* 187 (1934): 1–45; V. Volterra, "Variazione e fluttuazione del numero d'individui in specie animali conviventi," *Mem. Accad. Naz. dei Lincei* 2 (1926): 31–113.
36. L. Slobodkin, "Limits to Biodiversity (Species Packing)," in S. Levin, ed., *Encyclopedia of Biodiversity,* vol. 3 (New York: Academic Press, 2001), 729–38.
37. Volterra, "Variazione e fluttuazione."

38. J. Roughgarden, "Species Packing and Competition Function with Illustrations from Coral-Reef Fish," *Theoretical Population Biology* 5 (1974): 163–86; P. Yodzis, *Introduction to Theoretical Ecology* (New York: Harper and Row, 1989).
39. H. R. Akcakaya et al., "The Theory of Population Dynamics. II. Physiological Delays," *Bulletin of Mathematical Biology* 50 (1988): 503–15; E. Hernandez-Garcia et al., "Spatiotemporal Chaos, Localized Structures and Synchronization in the Vector Complex Ginzburg-Landau Equation," *International Journal of Bifurcation and Chaos* 9 (1999): 2257–64.
40. A. A. Berryman, "Intuition and the Logistic Equation," *Trends in Ecology and Evolution* 7: (1992): 316.
41. Slobodkin, "Population Dynamics in *Daphnia obtusa Kurz*"; U. Ritte, "Floating and Sexuality in Laboratory Populations of *Hydra littoralis*," Ph.D. thesis, University of Michigan, 1969; T. C. Griffing, "Dynamics and Energetics of Populations of Brown Hydra," Ph.D. thesis, University of Michigan, 1965; L. Slobodkin, "Experimental Populations of Hydrida," British Ecological Society Jubilee Symposium, *Journal of Animal Ecology* 33 supp. (1964): 131–48; G. Y. Yan et al., "Effects of a Tapeworm Parasite on the Competition of Tribolium Beetles," *Ecology* 79 (1998): 1093–103.
42. D. Dykhuizen, "Selection for Tryptophan Auxotrophs of *Escherichia coli* in Glucose-Limited Chemostats as a Test of Energy-Conservation Hypothesis of Evolution," *Evolution* 32 (1978): 125–50; M. Pantasticocaldas et al., "Population-Dynamics of Bacteriophage and *Bacillus subtilis* in Soil," *Ecology* 73 (1992): 1888–1902.
43. A. Crombie, "On Competition Between Different Species of Graminivorous Insects," *Proceedings of the Royal Society of London Biological Sciences* 132 (1945): 362–95; A. Crombie, "Further Experiments on Insect Competition," *Proceedings of the Royal Philosophical Society of London, Series B* 133 (1946): 76–109.
44. R. Strecker and J. Emlen, "Regulatory Mechanisms in House-Mouse Populations: The Effect of Limited Food Supply on a Confined Population," *Ecology* 34 (1953): 375–85.
45. C.R.B. Boake and M. J. Wade, "Populations of the Red Flour Bee-

tle *Tribolium castaneum* (Coleoptera, Tenebrionidae) Differ in Their Sensitivity to Aggregation Pheromones," *Environmental Entomology* 13 (1984): 1182–5.

46. T. Mahmood, M. S. Ahmad, and H. Ahmad, "Dispersion of Stored Grain Insect Pests in a Wheat-Filled Silo," *International Journal of Pest Management* 42 (1996): 321–24.
47. A. Łomnicki and L. Slobodkin, "Floating in *Hydra littoralis*," *Ecology* 47 (1966): 881–89.
48. N. Hairston Sr., *Ecological Experiments: Purpose, Design and Execution* (Cambridge: Cambridge University Press, 1991).
49. Slobodkin, "Population Dynamics in *Daphnia obtusa Kurz*."
50. G. F. Gause, *The Struggle for Existence* (Baltimore: Williams and Wilkins, 1934).
51. Slobodkin, "Experimental Populations of *Hydrida*."
52. R. T. Paine, "Food Web Diversity and Species Diversity," *American Naturalist* 100 (1966): 65–75.
53. Burgman and Lindenmayer, *Conservation Biology for the Australian Environment;* S. Pimm, *The Balance of Nature? Ecological Issues in the Conservation of Species and Communities* (Chicago: University of Chicago Press, 1991).
54. P. A. Colinvaux, *Why Big Fierce Animals Are Rare: An Ecologist's Perspective* (Princeton: Princeton University Press, 1978).
55. P. R. Ehrlich and A. H. Ehrlich, *Extinction: The Causes and Consequences of the Disappearance of Species* (New York: Random House, 1981).
56. L. Slobodkin, "The Good, the Bad and the Reified," *Evolutionary Ecology Research* 12 (2001): 1–14.
57. G. C. Daily et al., "Ecosystem Services: Benefits Supplied to Human Societies by Natural Ecosystems," in *Report No. 2* (Washington, D.C.: Ecological Society of America, 1997); F. Bazzaz et al., "Ecological Science and the Human Predicament," *Science* 282 (1998): 879.
58. S. Temple, "Plant-Animal Mutualism: Coevolution with Dodo Leads to Near Extinction of Plant," *Science* 197 (1977): 4–7.
59. D. Quammen, *The Song of the Dodo: Island Biogeography in an Age of Extinctions* (New York: Scribner's, 1996).

60. N. Hairston, F. Smith, and L. Slobodkin, "Community Structure, Population Control and Competition," *American Naturalist* 94 (1960): 421–25.
61. M. Davis, "Climatic Instability, Time Lags, and Community Disequilibrium," in J. Diamond and T. Case, eds., *Community Ecology* (New York: Harper and Row, 1986), 269–84.
62. R. Whittaker, *Communities and Ecosystems* (New York: Macmillan, 1970).
63. R. MacArthur, *Geographical Ecology* (New York: Harper and Row, 1972).
64. D. Botkin, *Discordant Harmonies* (New York: Oxford University Press, 1990).
65. S. Ferson, "Are Vegetation Communities Stable Assemblages?" Ph.D. thesis, State Unviversity of New York at Stony Brook, 1988.
66. Pimm, *The Balance of Nature;* Bazzaz et al., "Ecological Science and the Human Predicament."
67. Clinton, executive order on invasive species.
68. U. Safriel and Y. Lipkin, "Patterns of Colonization of Eastern Mediterranean Intertidal Zone by Red Sea Immigrants," *Journal of Ecology* 63 (1975): 61–63; T. Felsenburg and U. Safriel, "Colonization of Eastern Mediterranean Intertidal Zone by Indo-Pacific Mussel, *Brachidontes variabilis*," *Israel Journal of Zoology* 23 (1974): 212–13.
69. I. Rubinoff, "Central American Sea-Level Canal—Possible Biological Effects—An Opportunity for Greatest Biological Experiment in Man's History May Not Be Exploited," *Science* 161 (1968): 857–59.
70. Burgman and Lindenmayer, *Conservation Biology for the Australian Environment.*
71. P. Milberg and T. Tyrberg, "Naive Birds and Noble Savages—A Review of Man-Caused Prehistoric Extinctions of Island Birds," *Ecography* 16 (1993): 229–50.
72. Burgman and Lindenmayer, *Conservation Biology for the Australian Environment;* Pimm, *The Balance of Nature.*
73. L. Slobodkin and P. Bossert, "The Freshwater Cnidaria—or Coe-

lenterates," in D. Thorp and D. Covitch, eds., *Ecology and Classification of North American Freshwater Invertebrates*, 2nd ed. (San Diego: Academic Press, 2001): 135–154.

74. J. Beatty, "A Record of the Terrestrial Flatworm *Bipalium kewense* (Turbellaria: Bipalidae) from West-Central Texas," *Texas Journal of Science* 51 (1999): 105–6; P. M. Johns, B. Boag, and G. W. Yeates, "Observations on the Geographic Distribution of Flatworms (Turbellaria: Rhynchodemidae, Bipaliidae, Geoplanidae) in New Zealand," *Pedobiologia New Zealand* 42 (1998): 469–76; P. M. Johns, "The New Zealand Terrestrial Flatworms: A 1997–98 Perspective," *Pedobiologia New Zealand* 42 (1998): 464–68; P. K. Ducey et al., "Lumbricid Prey and Potential Herpetofaunal Predators of the Invading Terrestrial Flatworm *Bipalium adventitium* (Turbellaria: Tricladida: Terricola)," *American Midland Naturalist* 141 (1999): 305–14; J. J. Daly, H. E. Farris, and H. M. Matthews, "Pseudo-Parasitism of Dogs and Cats by Land Planarian, *Bipalium kewense*," *Veterinary Medicine & Small Animal Clinician* 71 (1976): 1540–2; H. K. Mienis, "Land Planarian *Bipalium kewense* in Israel," *Israel Journal of Zoology* 25 (1976): 71.
75. R. E. Ogren, "Predation Behavior of Land Planarians," *Hydrobiologia* 305 (1995): 105–11
76. Slobodkin, "Limits to Biodiversity"; A. P. Kinzig et al., "Limiting Similarity, Species Packing, and System Stability for Hierarchical Competition-Colonization Models," *American Naturalist* 153 (1999): 371–83; J. Roughgarden and M. Feldman, "Species Packing and Predation Pressure," *Ecology* 56 (1975): 489–92.
77. B. Mullin, "The Biology and Management of Purple Loosestrife (*Lythrum saliceria*)," *Weed Technology* 12 (1998): 397–401; M. G. Anderson, "Interactions Between *Lythrum salicaria* and Native Organisms: A Critical Review," *Environmental Management* 19 (1995): 225–31; M. Treberg and B. Husband, "Relationship Between the Abundance of *Lythrum salicaria* (Purple Loosestrife) and Plant Species Richness Along the Bar River, Canada," *Wetlands* 19 (1999): 118–25.
78. G. M. Ruiz and P. Fofonoff, "Non-indigenous Species as Stressors

in Estuarine and Marine Communities: Assessing Invasion Impacts and Interactions," *Limnology and Oceanography* 44 (1999): 950–72.

79. J. Levine and C. D'Antonio, "Elton Revisited: A Review of Evidence Linking Diversity and Invasibility," *Oikos* 87 (1999): 15–26.
80. E. S. Zavaleta, R. J. Hobbs, and H. A. Mooney, "Viewing Invasive Species Removal in a Whole-System Context," *Trends in Ecology & Evolution* 16 (2001): 454–59.
81. M. Sagoff, "What's Wrong with Alien Species?" manuscript, 1999.
82. M. L. Rosenzweig, "The Four Questions: What Does the Introduction of Exotic Species Do to Diversity?" *Evolutionary Ecology Research* 3 (2001): 361–67.
83. J. G. Vos et al., "Health Effects of Endocrine-Disrupting Chemicals on Wildlife, with Special Reference to the European Situation," *Critical Reviews in Toxicology* 30 (2000): 71–133.
84. J. D. Corser et al., "Recovery of a Cliff-Nesting Peregrine Falcon, *Falco peregrinus,* Population in Northern New York and New England, 1984–1996," *Canadian Field-Naturalist* 113 (1999): 472–80; D. E. Driscoll et al., "Status of Nesting Bald Eagles in Arizona," *Journal of Raptor Research* 33 (1999): 218–26.
85. R. E. Green, "Long-Term Decline in the Thickness of Eggshells of Thrushes, *Turdus* spp., in Britain," *Proceedings of the Royal Society of London, Series B—Biological Sciences* 265 (1998): 679–84.

Chapter 3: Two Major Current Problems

1. Office of Technology Assessment, U.S. Congress, *Preparing for an Uncertain Climate* (Washington, D.C.: Government Printing Office, 1993); Office of Technology Assessment, U.S. Congress, *Changing by Degrees: Steps to Reduce Greenhouse Gases* (Washington, D.C.: Government Printing Office, 1991).
2. R. A. Armstrong et al., "A New, Mechanistic Model of Organic Carbon Fluxes in the Ocean Based on the Quantitative Association of POC with Ballast Minerals," *Deep Sea Research II* 7 (2001): 1–18; R. A. Armstrong and R. A. Jahnke, "Decoupling Surface Production from Deep Remineralization and Benthic Deposition: The Role of Mineral Ballasts," *U.S. JGOFS News* 11 (2001): 1–2.

3. J. E. Cohen, *How Many People Can the Earth Support?* (New York: W. W. Norton, 1995).
4. R. Carroll et al. Strengthening the use of science in achieving the goals of the Endangered Species Act. Ab assessment by the Ecological Society of America. Ecological Applications 6 (1997): 1–11.
5. R. Carson, *Silent Spring* (Boston: Houghton Mifflin, 1962).
6. M. Sagoff, "Muddle or Muddle Through? Takings Jurisprudence Meets the Endangered Species Act," *William and Mary Law Review* 38 (1997): 825–993 (991).
7. A. Leopold, *A Sand County Almanac: With Other Essays on Conservation from Round River* (New York: Oxford University Press, 1949), 224.

Chapter 4: Applying Ecology

1. G. J. Hardin, *Filters Against Folly: How to Survive Despite Economists, Ecologists, and the Merely Eloquent* (New York: Viking, 1985).
2. Office of Technology Assessment, U. S. Congress, *Changing by Degrees: Steps to Reduce Greenhouse Gases* (Washington, D.C.: Government Printing Office, 1991); Office of Technology Assessment, U. S. Congress, *Preparing for an Uncertain Climate* (Washington, D.C.: Government Printing Office, 1993).
3. Adapted from a letter by Prof. Richard Schwartz, writing on January 7, 2002 to Coalition on the Environment and Jewish Life (COEJL), a Jewish ecological and environmental group.
4. B. Franklin, *The Autobiography of Benjamin Franklin* (New York: Black's Readers Service, 1932).
5. E. Gibbons, *Stalking the Wild Asparagus* (New York: D. McKay, 1962).
6. Excerpt of undated fund-raising letter from Gene Likens, director of the Institute of Ecosystem Studies, received December 19, 2001.
7. P. J. Hauptman and R. A. Kelly, "Digitalis," *Circulation* 99 (1999): 1265–70; R. Brustbauer and C. Wenisch, "Bradycardic Atrial Fibrillation After Drinking Herbal Tea," *Deutsche Medizinische Wochenschrift* 122 (1997): 930–32.
8. M. Maimonides, *Mishneh Torah* [Hebrew] (Jerusalem: Orekh Marav Kuk, 1964).

9. T. Browne, *Religio Medici* (Cambridge: Cambridge University Press, 1963).
10. H. M. Pachter, *Paracelsus: Magic into Science* (New York: Schuman, 1951).
11. L. Slobodkin, "The Good, the Bad and the Reified," *Evolutionary Ecology Research* 12 (2001): 1–14.
12. Quoted from a letter by Dr. David Slobodkin.
13. J. E. Cohen, *How Many People Can the Earth Support?* (New York: W. W. Norton, 1995); R. A. Rappaport, "The Sacred in Human Evolution," *Annual Review of Ecology and Systematics* 2 (1971): 23–44.

Conclusions

1. B. Wallace, *The Environment: As I See It, Science Is Not Enough* (Elkhorn, W. Va.: Elkhorn Press, 1998).
2. C. Elton, *Animal Ecology* (Oxford, 1927).
3. L. Slobodkin and A. Rapoport, "An Optimal Strategy of Evolution," *Quarterly Review of Biology* 49 (1974): 181–200; L. Slobodkin, "The Strategy of Evolution," *American Scientist* 52 (1964): 342–57.
4. R. M. Nesse and G. C. Williams: *Why We Get Sick: The New Science of Darwinian Medicine* (New York: Vintage Books, 1994).
5. M. Sagoff, "Muddle or Muddle Through? Takings Jurisprudence Meets the Endangered Species Act," *William and Mary Law Review* 38 (1997): 825–993.
6. L. Slobodkin, "The Peculiar Evolutionary Strategy of Man," in R. Cohen, ed., *Epistemology, Methodology and the Social Sciences: Boston Studies in Philosophy of Science,* vol. 71 (Dordrecht: D. Reidel, 1978), 227–48.

Index